KVHAA HANDLINGAR *Historiska serien* 24

Folke Dovring in the early 1960s, professor in the United States.

The Dovring Saga

A story of academic immigration

By *Janken Myrdal*

with contributions by
Gabriel Söderberg

KUNGL. VITTERHETS
HISTORIE OCH ANTIKVITETS AKADEMIEN
HANDLINGAR

Historiska serien 24

Myrdal, Janken. 2009. *The Dovring Saga. A story of academic immigration.* With contributions by Gabriel Söderberg. Kungl. Vitterhets Historie och Antikvitets Akademien (KVHAA), *Handlingar, Historiska serien* 24.
Stockholm 2010. 253 pp.

Abstract
Folke Dovring was a medieval historian in Sweden who tried around 1950 to introduce new ideas about history based on statistical analysis and history from below, publishing two books on medieval agrarian history that still remain important. His attempts, however, were made in vain. In 1953 he left Sweden for an international career, devoting himself to the study of the agrarian history of Europe in the 20th century, and in 1960 he became a tenured professor of Agricultural Economics at the University of Illinois, U.S.A.

Some major societal processes can be studied through the biography of Dovring. One involves European conservatism in the academic world. The second important process Dovring took part in was the expansion of American higher education. Dovring was one of the many scholars who left Europe in the 1950s to join the dynamic American academic community. When Dovring later shifted his focus from agrarian history to take up more political questions in the 1970s such as the oil crisis and environmental care his position became more isolated.

Key words
Agrarian history, history of the Middle Ages, academic conflict, higher education in the U.S., University of Illinois, land economics, waste of resources.

ISBN 978-91-7402-395-4
ISSN 0083-6788

Publisher	Kungl. Vitterhets Historie och Antikvitets Akademien, (KVHAA, The Royal Swedish Academy of Letters, History and Antiquities), Box 5622, SE-114 86 Stockholm, Sweden, http://www.vitterhetsakad.se
Distribution	eddy.se ab, Box 1310, SE-621 24 Visby, Sweden, http://vitterhetsakad.bokorder.se
Cover photo	Folke and Karin Dovring's home in Urbana, Illinois (1960–)
Cover Design	Lars Paulsrud
Graphic Design	Bitte Granlund, Happy Book
Photos	Private (Karin Dovring)
Print	Motala Grafiska AB, Motala, Sweden 2010

Contents

Preface

The Dovring Saga tells a tale that could be repeated countless times in the history of American higher education of scholars excluded, blocked, or intimidated in their native lands who made their way to the United States and found a place somewhere in its vast network of colleges and universities. This experience in turn is only one chapter in the larger history of a new nation peopled by immigrants who were oppressed or simply unable to make a living in their own countries and sought a new life across the ocean. Immigrants have always contributed mightily to the United States. In the process, they have also been participants in what may well be America's greatest achievement, its unending determination to try to assimilate all groups – ethnic, racial, religious, or otherwise – into a single nation with common loyalties and ideals. This is a process that continues to this day, confronting perhaps its greatest challenges in the ongoing struggle to give proper place to its large African-American population, descendants from the most shameful chapter in our history, and to respond appropriately to the massive influx of illegal immigrants who slip undetected across our borders, as they do in so many other countries today.

Professor Myrdal gives a flattering account of American higher education, together with a thoughtful summary of the various reasons why its universities have flourished during the last century. As he points out, one of the most prominent reasons for our success has been our good fortune in attracting so many talented scholars, such as Folke Dovring, who could no longer pursue an academic career in their own country. In a land with as many universities as the United States, it was unlikely that any promising scholar could fail to find a place or be blocked from further progress, as could happen in a country with a much smaller higher education system. Favored by the growing use of English as the lingua franca of the academic world, American universities were a natural home for individuals like Dovring who were seeking a new start in their careers. Since the United States never seemed to produce a surplus of exceptional scholarly and scientific talent from its own population, this influx of foreign intellectuals was a great boon.

As Myrdal points out, the sheer size and variety of American higher education offered other advantages as well. Universities could be found to suit any need or purpose. They could be private or public, large or small, secular or church-affiliated, single-sex or coeducational, vocational or exclusively devoted to the liberal arts, free-standing colleges or vast complexes with a full complement of professional schools. Such a varied array fitted the multiple needs of a huge, heterogeneous population. It fostered a competition that improved the quality of faculties, nurtured innovation, and encouraged vigorous efforts to attract new financial resources wherever they could be found. It offered far greater flexibility than was possible in the more centralized, state-controlled systems of Europe and thus could adapt far more easily to the tidal shift from elite to mass higher education.

At the same time, it only fair that I, as a representative of this system, should acknowledge its failings along with its strengths. Such a large system is unusually variegated and flexible, but it allows many institutions to survive at levels of quality that would probably not be tolerated in the more tightly controlled systems of many European countries. Moreover, if America's system is successful, its secondary schools are not, with the result that American colleges have had to spend inordinate amounts of time offering remedial instruction of a kind that should properly have been supplied in high schools. Finally, despite its many virtues, the spirited competition so characteristic of American universities has its dark side as well. Much money is wasted bidding for the services of professors from other institutions and moving them and their laboratories and libraries from one university to another. Competition has also spilled over into other activities such as athletics, sometimes with decidedly ill effects. Intercollegiate football and basketball have turned into a giant entertainment industry that leads universities to compromise their academic standards in deplorable ways by recruiting students with few if any talents or interests beyond their skills on the athletic fields.

One of the heartening features of higher education today throughout the developed world is a growing recognition of the vital contribution of universities to the welfare of their countries and a consequent desire to support their work and improve their quality. As a result, we have reason to hope that every country, not least the United States, will have something to learn from the university systems of other nations and that professors and students alike will increasingly teach and study in more than one nation. In time, in addition to welcoming foreign scholars like Folke Dovring, America may find its own professors gravitating to foreign universities, not only for a visit but for longer periods of time. The result of such a process cannot help but be beneficial for teaching, scholarship, and greater understanding throughout the world.

Derek Bok
President Emeritus of Harvard University

Introduction

There was a man, called Folke. His father was Ossiannilsson. They lived in Sweden. Then Folke married Karin. He fought at home, was defeated and went to America.

An Icelandic Saga often starts like this, and this is the Saga of Folke. In the Sagas persons appear through their deeds. In this biography about a Western intellectual, the hero appears mainly through his books and articles. This biography is more "letters" than "life" (to allude to another genre, the solid "British" biography).

Every life reflects the history of its period and will reveal something about society as a whole. The reader will find a similar use of a life-story here. The subject is Folke Dovring, medieval historian in Sweden and professor of agricultural economics in the U.S.A. at the University of Illinois, Urbana-Champaign. Besides telling trajectories in his life, this book is also about academic hierarchy and intrigue. Furthermore it gives homage to that great country in the west – or rather to some of the streams of peoples, ideas, and events that made it great.

What about Folke's life-story as such? A dark thread of his fate can be traced from his early youth: well-meaning but overly ambitious parents who pushed their boys. The youngest boy, Folke, was gifted and aware of the fact, but he had difficulty garnering recognition from his father. Another thread in Folke's yarn of destiny became intertwined with the first, as he, young and bold entered a conservative fraternity of historians: fighting internal battles. The young man tried to introduce new ideas about history based on statistical analysis rather than speculations about the actions of kings and higher nobility. His attempt was made in vain, and eventually he was rejected by the Swedish academic community.

After leaving Sweden his international career lead him to the U.S.A., at a time when this country was taking the leadership position in the international society of academics. For many years he lived as a professor of a renowned university, in a small town in the heartland of the United States. His political interests intensified. Viewing America through Swedish eyes, he wanted to promote ideas about equality and protection of the environment; ideas not mainstream in America at that time. Again he assumed an underdog-position.

How it all started

I am an agrarian historian, and when I worked on my dissertation about medieval agriculture in Sweden, I encountered the name Folke Dovring. He was the author of two major books on tax systems and landholding in the Middle Ages, published 1947 and 1951. I read them very thoroughly, taking copious notes, and referred to them in my dissertation. Later I realized that Folke Dovring had been the first agrarian historian of significance in Sweden. He was appointed assistant professor ("docent") of agrarian history, and in 1953 wrote a textbook called *Agrarhistoria* (Agrarian History). A note on terminology is that "agricultural history" is about the history of agriculture as such, whereas "agrarian history" also includes the context surrounding agriculture such as social structure and household organization.

I was also vaguely aware of an academic dispute involving him and that he no longer lived in Sweden. His works were used but he was never mentioned as a person among the medievalists in Sweden.

Eventually, in the mid 1990s, I was appointed to a chair in agrarian history at the University of Agricultural Sciences in Uppsala. There had been no full professor in that field in Sweden before so I started to build a unit for research and education. I assigned Dovring's textbook from 1953 to my graduate students to illustrate how far the field had developed half a century ago, but I also began to wonder – what had happened to Folke Dovring? I knew that in 1956 he had published a major book about agriculture in Europe during the first half of the twentieth century, but then? A friend of mine, returning from a conference, told me that Folke Dovring was alive and living in the United States. He gave me an address. With the help of directory assistance I could thus find a phone number. I hesitated. Should I phone him, and what would I say? I started to pursue funding to invite him to Sweden, and was ultimately successful. After a couple of months I dialed his number. A woman answered. I asked for Professor Dovring. Sobbing she replied: "He is dead. He died suddenly two weeks ago."

That is the moment this investigation began. Questions arose and demanded an answer. Why had there been such silence about Dovring in Sweden? Why did he leave when he had written those marvelous books? Why did he then turn up in the U.S. after initially publishing on the Middle Ages in Sweden and then on the early twentieth century in Europe? In Sweden the breakthrough for agrarian history was delayed by several decades upon Dovring's departure. In the Netherlands, in Germany, in England and in many other countries the history of agriculture and agrarian history became an established discipline after World War II.

When, as a result of this contact, I was invited to the University of Illinois, I had the opportunity to answer these questions.

Type of biography

"Is Dovring worth a biography?" was one of the first questions I encountered when I came to the University of Illinois to write about one of its former professors. The question was asked by one of the historians at the university, having no knowledge of either Dovring's career as an economist or as a medievalist. This professor also was of the not uncommon opinion that biographies should be reserved for the most famous and influential.[1]

Unprepared for the question, I did not formulate a clever answer at that moment. My answer to the question could have been that any life is of interest: that of a farmer, a worker or an intellectual. I do not pretend that Dovring was the most significant scholar of his generation, but he had a certain influence on medieval studies in Sweden and on agrarian studies in Europe. Later he worked with the burning issues of his time: land reform, energy crisis, the affluent society. He often took a position outside the mainstream of ideas.

A metaphor can be helpful. When looking at a fountain I often think of humanity as a whole. At the top we have separate drops dancing for a short while in the air, as if they where flying, yet they also fall and disappear. At the bottom others drop out early to be swallowed by the pond. In the middle we have a strong jet of water where the separate drops cannot be seen, but they will all in time fall back to the water from where they came. Significantly, the drops at the top owe their existence to the jet of water, throwing them up. Those dancing in the air first catch our eye, then our gaze wanders downwards and we grasp the beauty of the whole fountain.

When reading biographies of scholars with a moderate impact you often find an excuse in the beginning, explaining that such scholars also are worthy of a biography, and this biography so far seems to be no exception to this rule. But in my opinion such an excuse is actually not necessary. Therefore I instead turn the interest to which type of biography is at hand.

Biographies can be written in at least four different ways:

1. *The ego*. This approach concentrates on the narrative, and often includes a psychological portrait. The biographer tries to delve into what is secret and concealed, such as lingering madness or unconventional sexual habits. This style of writing is sometimes close to fiction for much of it is guesswork. In this type of biography the reader can more easily identify with, or at least relate to, the subject. If the subject is famous and held in high esteem, such books can be written again and again on his or her life, with new interpretations and revisions.

2. *In the world*. A second way to write a biography is to let the life-story illustrate an epoch. This is often the case when the material is so thin that it only offers fragmentary glimpses. Thus this kind of biography is often found in books about medieval history. Such biographies are also common about ordinary, not famous people who the public cannot be expected to have an interest in as such, but who enable contemporary history to stand out more clearly through their individual fate.

3. *Influence by*, being influenced by predecessors. A specific type of investigation about a person is to look at the work they produced, and in which their intellectual environment took shape. This is typical in the history of literature and art. The scholar writing on an author, artist or the like often asks: who influenced the

1 Compare Hankins (1979), p. 11.

subject, can strings of influence from predecessors be identified in his or her production?

4. *Influence on*, exerting influence on his or her heirs. A similar question, still looking at the work he or she produced, but from the opposite position, is to ask: who was influenced by the subject of study? In a history of philosophy the study often involves understanding the meaning of the text; the subject's influence on later writers follows as a consequence of comments on the textual source of information. This approach is actually the type of biography that most subjects would like to read about themselves – what imprint have I made on the world, does anything of me survive?

These four approaches merge into each other, but I will mainly use the second and the fourth way of writing a biography. Defending the use of biography in the history of science and scholarship, Thomas Hankins declares that it is often difficult to bring personal life together with the subject's scientific work in a harmonious way.[2] I can only agree.

I will not talk much about his personal life after youth, as it did not in any obvious way influence his scientific work. Folke Dovring was a vegetarian, and is said to have been a good cook, but I do not intend to talk about the food he liked or disliked. He also was fond of gardening and tried to shape a Swedish-like garden on the deep soils of Illinois, a nearly hopeless project, which I am not going to discuss further. He and Karin never had children, much because of their vagrant life until they were nearly forty, but he was fond of children and when established in the U.S., they pondered about adopting a child, but this thought was not realized. Folke loved cats, and had the specific gift of talking with cats. I can do that but I will not reveal the secret in this book.

In an introduction to biography-writing Marc Pachter warned against "the voluminous compendia of facts-shovelled-on-facts in which the biographer buries alive both his hero and reader".[3] He argued that a biographer must catch the essence of a life and make it convincing. A biography about a scholar must take the scholarship of that person seriously and see how he or she went about his or her task, how ideas developed, and how they were tested.[4] Surprisingly often biographers of scholars do not take a standpoint on how accurate or important the scientific results produced by the scholar were. Such an evaluation will change as time goes by, with new findings, but none the less I feel it must be done.

I certainly have no intention of writing a laundry-list-biography[5] or a deep psychological reconstruction. Instead this book is centered on a few basic questions – all related to Dovring as an academic.

2 Hankins (1979), p. 2–3.

3 Pachter (1979), p. 3.

4 Pachter (1979), p. 3–4.

5 Tuchman (1979), p. 146–147.

The questions

The first question of the book is: Why was a hard-working, innovative, internationally successful young scholar expelled from the academic community in his homeland? Basically, this question concerns human nature as it is evidenced in the university milieu. It is a question every scholar or scientist in a position of power ought to ponder regularly.

This question can be given a more specific twist, asking: What structures in the European university system after World War II, specifically in the Swedish one, made it difficult for innovative scholars to reach the top? I investigate the conservative structures, which could hinder a young historian even though he had produced groundbreaking books. Dovring advanced ideas that later mainly turned out to be correct despite being rejected by the leading professors of the time. Understanding impeding structures in order to avoid them for the good of science and humanities is imperative, and the Dovring-case is an important illustration of such structures.

The second question is founded on a claim: U.S. domination of the world is not only, perhaps not even primarily, based on economic and military power. In addition, cultural domination (through film and music) and intellectual hegemony (through the leading universities) play an important role. Two thirds of the world's leading universities are located in the U.S.A. (see Appendix 2), and the international academic community revolves around their activities. My question is: how did this come about?

Today we live in an epoch where data and information are produced and spread with an efficiency that directs the whole social structure. The world of academia is formed as one unit, with the U.S.A. at the top and other countries fit into an ensuing hierarchy. The 1950s and 1960s were decisive decades in the United States' takeover of the intellectual world arena. Dovring was one of many scholars who left Europe to join the dynamic American academic community, a pattern of intellectual immigration that was a major cause of its growth spurt. Dovring also contributed to this amazing expansion by mentoring a significant number of M.A./M.S. and Ph.D. students, many of them, like himself, natives of other countries. But we have to ask: why did a department in an American university decide to give Dovring a secure position as a tenured professor, and why did he choose to go to America?

In this case of intellectual immigration Dovring stands out as typical. He was also atypical, namely in his political ideas in the U.S.A. The atypical can reveal the typical, by making dominant structure palpable. This was what Dovring did against the society of Swedish historians, and again in America against the typical economists of the 1960s and 1970s. He developed his program out of love for his new homeland. However when he focused on U.S. dependence on the automobile and fossil fuel as the underlying problem, and declared that freeways were a mistake, and further recommended reducing the number of parking places in cities, he was marginalized. But he was one of the first to advocate non-oil based fuel, and also one of the first to point out that oil dependence would drag the U.S. into the quagmire of Middle Eastern politics. Again reactions reveal something about the American at the time.[6]

A person like Dovring causes the surrounding typical persons to react in public, which exposes their typicality. The third question is what a scientific underdog position can say about the economy in the U.S.A. after World War II. Dovring's ideas on preservation and politics can at least partly be seen as foresighted, and the way they were ignored is an interesting sign of the dominant paradigm.

Sources and methods

Four different sources have been used: texts, official documents, private letters and interviews.[7] The most important of these sources are published texts written by Dovring. Methods are mostly qualitative, including text analysis, discussing historiography and reviewing research on the expansion of the U.S. university system. Quantitative methods have mainly been used when I have tried to measure influence (presented in an appendix) or work effort.

The texts are all enumerated in the bibliography at the end of the book. Many of them are difficult to find in ordinary libraries. Most of them can of course be found at the University of Illinois.[8] A list of the graduate students he supervised is of importance for understanding his role at the university.

At the university library in Urbana several boxes contain "The Dovring Archive" (quoted as: DA). There most of the official documents concerning Dovring are preserved (of especial interest are those about academic conflicts).

A large collection of letters written to Dovring has been preserved in this archive, along with some copies of letters he wrote himself. From Dovring's relatives in Sweden I obtained a collection of letters he wrote, mainly to his mother. (Copies of these letters have been given to KSLA in Sweden.) I quote them by giving the date and the person sending or receiving the letter. In a few cases I have used other archives to find letters, or documents concerning Dovring.

I also have conducted interviews, both with Folke Dovring's former friends and colleagues in Sweden, and those in the U.S.A. (enumerated in an appendix).

---

6 See also Appendix 4 by Gabriel Söderberg.

7 I am aware that a more extensive search could have been done in different Swedish and American archives, not the least to find more letters from Dovring. My excuse is that the texts he wrote and published stand in the center of my investigation.

8 I have also given a collection of his books and articles to The Royal Academy of Agricultural Sciences in Sweden (KSLA, that is: Kungl. Skogs- och Lantbruksakademien), see the database LIBRIS.

A theme in this book is academic conflicts. They can be very thrilling for those involved but for others their interest lies more in what they reveal about the academic society and intellectual history. These conflicts are always grounded in the scientific debate; texts must constitute the kernel of an interpretation and a biographer must try to understand and identify the essence of the debate. Being a historian, this was less of a problem for me in investigating Dovring's research on the Middle Ages than his writings on economics.

Understanding the issue at stake is seldom enough; other factors such as prestige, authority and personal relations have a role. Furthermore, conflicts always have a hidden agenda as intrigue per definition, is made in the obscurity. Official documents about the academic promotion procedure give some hints, especially in Sweden where the state bureaucracy demanded and produced detailed dossiers. Letters and interviews fill out the picture, but letters are always targeted to the recipient. Writing to a friend is different from writing to an enemy, and it is not to be taken for granted that the letter to the friend gives a more "correct" description. Interviews are always affected by memory as a reworking of the past, but they can give information not available anywhere else.

Another theme is America's success story in academics. In this, I start with the numbers of students and teachers. Everything can be measured and nothing can be explained just with measures. We can answer questions of how much, when, and where, and we can compare the size of different elements. However this frame of knowledge has to be filled with details. When Dovring and his department are seen as a case study, I follow a path in his life leading him from an odyssey in Europe to his new homeland. This conjoins with a trajectory followed by the department at an American university where the leaders tried to break a tradition of inbred thought and open the doors to competition and change.

Perhaps I can be accused of being obsessed with measuring impact, but, if so, I am only reflecting an obsession in the system of academia. Dovring himself also had such an obsession, and he had quite a collection of *Who's Who* in which he was mentioned. A reason for this obsession is competition, which demands comparativeness and this leads to measures. No single measure gives the whole picture. Several are tested to give a numerical frame of reference to a discussion about why Dovring was, or was not, accepted by peers and why he did or did not reach out to a larger audience.

A frame of a scholar's life

Measures give a frame for a deeper understanding; therefore I here present a frame. I will try to describe his work, especially his publishing activity, which was not only the part of his effort that he himself held in highest esteem but also the part of his life in which I am most interested.

Often, especially in natural science, the number of articles published is used to measure the output of scientific activity, but in humanities and social science the number of pages is a better measure. Books and articles can be anything between ten pages and five hundred pages. Normally, a longer text requires more work, especially if we are analyzing a scrupulous scholar, which Dovring undoubtedly was. Thus the number of pages gives a rough idea of where work was invested as presented in figure 1.

Any intellectual has to plan his or her life several years in advance. A project usually takes many years to accomplish, and results are only seen after years of work. Thus the curve in figure 1 should be adjusted to the left, if one wants to study his lifetime writing project.

I later will use the same method, measuring the number of pages, when I discuss his main shift of interest from the Middle Ages to the contemporary period; from Sweden and Europe to the U.S. and the world. The number of pages in different categories shows the change.

Dovring's bibliography lists more than 200 publications, a total of 6 000 pages. I have presented the total amount of pages per year in figure 1. There are no quality aspects in the diagram, separating different kinds of publications from each other. Pages in books published by prestigious publishers carry the same weight as mimeographed leaflets published by the department.

In one glance at the figure we get an overall view of Dovring's life. He had an early peak around 1950 when his dissertation and ensuing books came out. During his years as an administrator in Rome, in the late 1950s, he experienced a decrease in the flow of publications. His first decade in the U.S. was marked by more intense writing and publishing, and thereafter he had a fairly even produc-

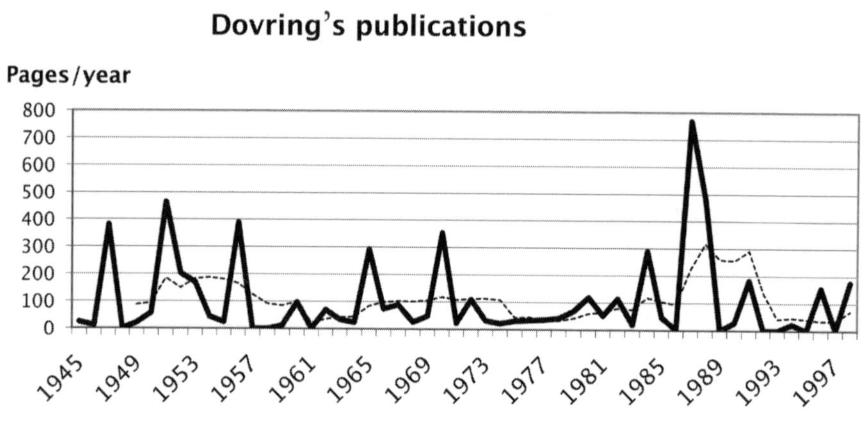

Figure 1. Dovring's publications, number of pages per year (dotted line is the moving five-year arithmetic average).

tion during his years as a tenured professor in Illinois. Before and after retirement he had an upsurge of publishing, with several large books.

Basic dates

Some basic dates in Dovring's life give another framework:

He was born December 6, 1916;
Graduated from high school, 1935;
Did his military service at the field artillery with geodesic service, 1936–37;
Received his B.A., 1939, at Lund University;
Worked as a book auctioneer and at the university library in Lund during the years before presenting his Ph.D.;
Served in the army for several periods during the war;
Married Karin Engström, 1943;
Earned his Ph.D., 1947;
Was assistant professor (*docent*) in Lund, 1947–1953;
Had a Rockefeller fellowship in Geneva and Rome, 1953–1954;
Worked at FAO in Rome, 1954–1960;
Became a tenured professor, University of Illinois, Urbana-Champaign, U.S., 1960;
Became an American citizen, 1968;
Retired from the University of Illinois, 1988;
He died May 21, 1998.

Ethical questions

When I started this project, I believed writing biographies to be an easy task. Usually I cope with old history, often medieval, where a few pieces of information must be scrutinized to give information. Now there seemed to be plenty of sources and even living people with whom to converse. I had not realized that biography in some respects is the most difficult of all branches of historical analysis. It is close to fiction in the telling of a story, and, at the same time, has a very strong emphasis on non-fictional correctness, as friends and relatives are among the readers.

I had to contend with the moral and ethical problems of writing about a person who had lived in my own time. As every other biographer, I encountered family conflicts where time has not healed any wounds. A colleague, who had written biographies, gave me the following advice: never talk about anything unpleasant. Of course I could not follow his advice. Jørgen Sandemose, the son of the famous Danish-Norwegian author Aksel Sandemose, wrote a biography of his father. In the ensuing discussion about his book he formulated the following sentence: the importance is not to differ between what ought to be private and what is of public interest; but to identify what ought to be public and what only is of private interest.[9]

Acknowledgments

I never met Folke Dovring. Some of my questions would have been put differently and also answered differently in this book, had he been alive. Many have helped me. First among those who have helped me stands the widow of Folke, Karin Dovring, a remarkable woman worthy of a biography of her own. Another relative of importance regarding this book is Folke's niece, Maj Sundblad in Uppsala, Sweden, who loaned me a series of letters from Folke Dovring. In addition there are a number of professors at the University of Illinois, neighbors in Champaign-Urbana, friends at home in Sweden and others who have contributed to my understanding of Folke. I especially want to mention professor Earl Swanson, professor Shelley Wright, who read and made corrections to the manuscript, and Marlene Mather who read the whole text twice and made several amendments. I have gotten support from the heads of the Department of Agricultural and Consumer Economics during the process of my work: Darrel Good, Andrew Isserman, Robert Hauser.

The manuscript was discussed in a seminar at Kungl. Skogs- och Lantbruksakademien (The Royal Academy of Agricultural Sciences) 2005 with comments from docent Britt Liljewall, dr Ronny Pettersson, chief librarian Lars Ljunggren and others. I also in 2007 gave a presentation at Kungl. Vitterhetskademien (The Royal Swedish Academy of Letters, History and Antiquities) and got comments from professor Birgitta Odén, professor Göran Dahlbäck and others.

These people have talked with me; shown me bits and pieces to put together – but the interpretation is all mine.

The manuscript was basically finished 2007 and for various reasons the publishing has been postponed. In finishing the manuscript, research assistants Gabriel Söderberg and Eva-Lotta Päiviö have been very helpful. Söderberg has written an extra appendix about Dovring as an economist, and also about Dovring's last book wherein he touched upon religion.

Without financial support from the University of Illinois and the Department of Agriculture and Consumer Economics at the same university, this study could not have been started or accomplished. I have spent a most contented time in the heartland of the U.S., during my efforts to understand a remarkable man, his time and milieu. Kungl. Gustaf Adolfs Akademien (The Royal Gustavus Adolphus Academy) has funded journeys to the U.S.A. from Sweden. Kungl. Vitterhetsakademien has decided to publish the book, for which I am very grateful.

9 Sandemose (2005), p. 21.

CHAPTER 1
Childhood and the father (1916–1934)

The father and mother

We repeat the past, but in new ways. The survival strategies and emotional responses we develop within our families are transferred to our work situation, and vice versa. All human beings absorb patterns from their parents and often, in one way or another, repeat aspects of their lives, whether deliberately or by chance.

I do not generally like the metaphor of life and society as a theater, since the acting we do is for real. We do not go home and remove our deeds as if they were costumes and grimaces. Love, work, friendship, and so on are genuine. But, because we repeat ourselves, I can nevertheless agree with Shakespeare: the world is a stage, and life is a drama we perform not only for those around us, but, perhaps even more, for ourselves. Strategies we develop as children are powerful and unconscious. Childhood counts. The person who dominated Folke Dovring's childhood was his father.

Karl Gustav Ossiannilsson was a famous author. Early in his career, at the beginning of the twentieth century, he attracted considerable attention, establishing himself as an up-and-coming writer. His reputation seemed secure. But destiny took another turn. Though Ossiannilsson continued to be prolific for many decades, publishing many well-crafted novels, his standing gradually slipped and he never achieved the renown predicted by his early success. Today Ossiannilsson is hardly read at all.[10] He was the father of Folke Dovring.

K.G. Ossiannilsson's life has mainly been described by his enemies, who have concentrated on the first years of intense activity. An exception is Dag Hedman, in an article in *Svenskt biografiskt lexikon* (the Swedish biographical dictionary), who provides a more balanced assessment.[11] The summary below draws mainly on Hedman's account.

Born in 1875 in southern Sweden, Karl Gustav Ossian Nilsson was the son of an organ-maker. He began writing poems and plays while a student at Lund University. When he started to publish, he changed his last name from the plain Nilsson, a common patronymic, to the more eye-catching combination of his third forename and his surname, calling himself K.G. Ossiannilsson. (Folke Dovring

10 Cf. Furuland (1989), p. 237.

11 Hedman (1994), p. 400–407. See also for a more panegyric description Engberg (2003). Neither of them discusses the family conflict or mentions Folke Dovring.

would later change his name at a similar transitional period of his life.)

Ossiannilsson's verse was powerful, fluid, and uplifting, with bold cadences and a rhetorical flourish. In his poems he often lauded the man of will, the strong personality who could change history. At the same time he sympathized with the oppressed and regarded those who brought happiness to the masses as heroes.

During the earliest years of the twentieth century he was deeply engaged in the young and fast-growing Social Democratic Party, composing political songs in its service. In 1908, however, he broke decisively with the Party by publishing a book (*Barbarskogen*, which is: The Barbarian Forest), which delineated the narrow mindedness and bickering he had encountered among its members.

After 1914, Ossiannilsson turned his attention to novels highlighting exciting and adventurous events in Swedish history. Appearing at the rate of one or two each year, many of these novels reached a wide readership. He also published short stories in one of Sweden's most important weekly magazines. Ossiannilsson's fiction may, in fact, have formed many Swedish children's view of their own history. Another lasting contribution was Sweden's national flag-anthem, still in circulation today.

However the leading critics, especially the influential Fredrik Böök, did not regard him highly and their opinion eventually won the day. They acknowledged his craftsmanship but also pointed to his lack of psychological insight. "Decorative but superficial" was their harsh judgment, one that came to be the standard comment about Ossiannilsson in reference works.

Ossiannilsson's approach to political questions became passé when the Swedish political landscape changed. Fierce fighting between the socialists and their adversaries was modulated and modified, leading instead to negotiation and compromise. He entered the fray once more when drums and flags again came to dominate politics. During the late 1920s and early 1930s he paid homage to Mussolini as one of the strong men of history, touring the country for many years proselytizing Italian fascist ideas and involving himself with far-right fringe groups. Later, however, when the war broke out, he disassociated himself from Hitler and the Axis, and, during the war, he supported England and her allies (admiring Churchill).

By that time, however, he no longer had any political influence. Though he wrote his autobiography immediately after World War II and was productive until late in life, his works went unnoticed by the cultural establishment. He died at the age of 95 in 1970.

Hedman notes a unifying characteristic in Ossiannilsson's activities and literary production: he wanted to educate, to elevate the public and the nation. He could also be described as opinionated, a trait that may help explain his most obvious weakness as a writer; his inability to create rounded, contradictory, and thus convincing fictional characters. As a political writer his hero-worship tended to exclude him from meaningful public debate.

Ossiannilsson married Naemi Arnman in 1905 when his literary career was on an upward trajectory. She was ambitious and intellectual, and, like him, had a university education, a relatively uncommon achievement for women at that time. She had worked as a high school teacher in the small provincial capital of Växjö. At that time secondary schools were gender separated, so she taught at a girls' school. When Naemi inherited an estate in central Sweden, the couple left the university milieu in Lund and started to build a house and a new life at Frösta in Östergötland in 1915. Between 1908 and 1916, Naemi gave birth to five boys. Folke Dovring was the youngest.

Raising five boys

The couple decided to home-school their sons. Both were teachers who regarded the boys as raw material for them to shape without interference from the public school system. The boys were also put to physical labor, helping build the house and cultivating the land. The family was a hard-working little community under the leadership of a patriarch. During the evenings, the father read aloud to the boys from the classics of world literature, always in the original language, German, French, or English. He had a resonant voice, much admired by others, and even in old age continued to enjoy reading aloud to his grandchildren.

The boys were encouraged to be competitive and not surprisingly, the oldest son hoped to go on winning forever, even when the age-related advantages of size and skill had evened out. This trauma, experienced by every first-born child, was intensified in this instance by the father's expectations. The parents seem to have referred to him as their favorite, the one they regarded as the future hope of the family.

The parents also began pondering what careers the boys should pursue as adults. For instance it was determined that the oldest would be a natural scientist, the second a military man, an officer.

Last came Folke. For many years he tried to avoid the homework his parents assigned, preferring instead to go on nature walks, and eventually becoming a skilled hunter. His parents decided that he should be a farmer or work in agriculture, or alternatively he could stay at home and help them when they grew older. Faced with this treat, he started to take more interest in his parents' endeavors to teach him, revealing considerable ability, particular in languages as well as other subjects.

Of the plans the parents drew up for their sons, only those for the oldest boy, Frej, were fulfilled. He became a scientist, though music was an ongoing serious hobby. His thesis, in fact, combined these interests, as his topic was the drumming sound made by one particular insect. The second boy, Sölve, dropped out of military training and later became a high school humanities teacher. The twins, Dag and Göran, died young, though Göran actually did embark on a career as an artist.

In Sweden young people normally do not enter the university until they are nineteen or twenty years old. First they must graduate from secondary school. Home-schooled youngsters could pass a special examination as private students before an official Board of Education in Stockholm. They usually were tested more rigorously than others since they had not demonstrated their abilities in the classroom. All the Ossiannilsson boys, however, did well and Folke, when he graduated in 1935, actually received one of the best scores in the entire country.

Three years before Folke matriculated, catastrophe struck the family: Dag drowned while out swimming with his oldest brother in a nearby lake. The father wrote a long poem about the loss of his son; the mother withdrew from social life and for many years dressed in black.

Another significant event at about the same time was the arrival of a new housekeeper, Vivi. She was a young girl, only two years older than Folke, and came from an impoverished farming family. As did all the servants, she lived in the house with the Ossiannilsson family, but unlike those who had come before her, she remained in the household, developing a close relationship with both parents.

Eventually Vivi became very close to K.G., and his wife Naemi most probably came under pressure. A few years later she experienced a mental breakdown. Causes for this breakdown are outside my investigation, but the new situation came to influence Folke's life. I will come back to that. In 1945, K.G. had his wife declared incompetent and committed to a mental hospital. She was never to return to her home, and remained hospitalized except for a period from early 1952 until early 1958 when she lived with her oldest son, then a professor at Uppsala. Shortly after Naemi's death in 1961, K.G. married Vivi, the housekeeper.

Folke's relationship with his mother was complicated. He felt close to her and kept in touch by letter during her periods of hospitalization. Apparently she had great hopes for her youngest son. In March 1947, when he had finished his doctoral thesis, he wrote to her that "your dreams and mine have come true: I really will be a scientist and a university teacher",[12] and the next letter informs her he has earned the doctoral degree that "I promised you 12 years ago, at my graduation from high school, I would receive."[13]

But this encouragement also created pressure, as Naemi demanded success and perfection from the entire family. Like many women of the period, she had no outlet for her own talents, so her ambition was expressed vicariously, through the men around her, the father and the sons.

Background to a family conflict

One of the major shifts in Western mentality during the last century is the gradual democratization of the family. Wives, younger siblings and sisters have

12 PA-MS, Folke Dovring to Naemi Ossiannilsson, 27.4.1947.

13 PA-MS, Folke Dovring to Naemi Ossiannilsson, 9.6.1947.

strengthened their position against husbands and older brothers. This has caused an enormous amount of family strife, but the change can be seen as a part of a larger process of democratization in our society. In the Ossiannilsson family Folke stood for the new values, against his father and oldest brother.

Under a patriarch there are only two possible strategies, to submit or to rebel. Folke chose the latter. Temperamentally, in this regard he more closely resembled his father than did the other brothers, even though K.G. never had the slightest understanding or sympathy for Folke's actions and motivations.

The conflict developed in stages. Seen from the outside with knowledge of all the emotions involved in every family conflict, one must accept that it is not possible to describe without writing a novel. Small but significant words loaded with hidden meaning, undercurrents, and minor skirmishes not registered in any source certainly had far-reaching effects.

As a student in Lund, Folke had met Karin. They formed a union, eventually as a married couple but always supporting each other emotionally and intellectually. In the conflict in his family she stood on Folke's side. On his father's side stood the rest of the family as well as other relatives. Often it takes more than a lifetime to resolve and heal a family conflict. Even fifty years afterwards, the memory of the strife arouses powerful emotions among those who were, to a greater or lesser degree, involved.

Folke initially wanted to be a forester, but for some reason decided to give up these plans despite his parents' encouragement. He was told by a teacher to try history. When he came to the university in Lund his two oldest brothers were already enrolled there, and all three maintained contact with each other. The second brother, Sölve, and Folke attended the same history classes during 1936.

He had other close friends as well. One of them was Torsten Husén, who also studied history in 1936. He told me that he, Folke and some of their friends used to visit the Ateneum coffee shop, world famous in Lund. They would sit over a cup of coffee talking for hours, as students always have and hopefully always will. Other close friends at this time were the brothers Nils and Per Stjernquist and Gerhard Hafström.

They were a gifted group of students. Per became a tenured professor of law in 1950, Nils a professor of political science in 1951, Torsten a professor of education in 1953 and Gerhard a professor of history of law in 1955. Folke was the last to become a tenured professor. Later on, Folke certainly remained aware of the careers of his friends, as they followed his.

In the devastating ideological and military conflict that tore the world apart from the late 1930s, families were split. The surviving twin, Göran (who was also to drown in an accident some years after the war), was the only family member with clear Nazi sympathies. The oldest son, Frej, supported the Allies. It seems likely that Folke agreed with Frej, at least dating from around 1940.

In a letter of 2 November 1941, Frej wrote to Folke congratulating him on his

Folke, a young man in the military service 1941.

Folke's parents. K.G. Ossiannilsson 1934 and Naemi Ossiannilsson in the early 1930s.

engagement to Karin, adding that he had come across an "interesting and highly enlightened journal, *Nordens frihet* [Freedom in Northern Europe]. Are you familiar with it?"[14] This journal was anti-fascist and anti-communist, supporting the Finns in their fight against the Soviet Union but also the English in their struggle. The letter seems to indicate that the oldest and the youngest brother were on cordial terms and held similar political opinions.

The first challenge

Folke Dovring cannot be understood unless his wife Karin is taken into account, since they lived in a symbiotic emotional and intellectual relationship for so many years.

Karin Engström and Folke met in the university library in Lund in 1939. She had heard of him as a person who, somewhat astoundedly, had completed the undergraduate curriculum in two years instead of the usual four or five. The first time she saw him he was talking animatedly, surrounded by other students. Though she found him a bit pretentious, she also admired him.

Not infrequently, it is the woman who takes the initiative, even if, out of courtesy, she lets the man believe otherwise. And so it was in this instance: the intelligent young woman saw the brilliant young man and made up her mind before he did.

Karin's family background was even more difficult than Folke's. She grew up with her grandmother, her mother having placed her with a foster family for a year, taken her back, and then chosen not to raise her daughter herself after all. From the age of fifteen, when her grandmother died, Karin lived on her own. Her mother had remarried and Karin was not accepted by the new family. She was, however, considered intellectually gifted and was awarded scholarships to excellent schools, and eventually to the university.

When Karin was introduced to the Ossiannilsson family, she was at first welcomed enthusiastically. She was the first (future) daughter-in-law with a higher education, which was highly esteemed by K.G. and Naemi.

Surviving correspondence from K.G. Ossiannilsson to Karin during this period reveals a friendly, even confidential tone. In a letter dated 5 March 1942 K.G. sends her one of his books, certain that she will enjoy it (unlike most people, he adds bitterly). Referring to German bombing raids, he laments the destruction and cultural loss.[15]

Evidently alluding to Karin's teasing him about a supposed new young mistress, he rejects her insinuation in a lighthearted tone. One really wonders if Karin dared to allude to the complicated family situation or if she was just innocent. K.G. concludes by praising her stance in favor of state support to families with

14 PA-MS, Frej Ossiannilsson to Folke Dovring, 2.11.1947.

15 PA-MS, Karl Gustav Ossiannilsson to Karin Dovring, 5.3.1942.

children, apparently a topic discussed when they had met.

During the summer of 1942 a break occurred, brought about when Folke accused his parents of treating him unfairly with regard to financial matters. One source of contention was the upcoming wedding of Karin and Folke, for which he felt his parents should pay. His father refused. Since no letters from Folke are extant, it is impossible to judge whether he deliberately provoked his parents or merely broached the question gingerly. His father, however, responded with a counter-attack.

In a letter of 10 December 1942, K.G. goes so far as to upbraid Folke for endangering his mother's life with his allegations, since she loves him (Folke) "more than you deserve."[16] K.G. continues, "Blame me instead, I can endure it" but then adds in his own defense, "Before you harass me, take a look in the taxpayers' directory. Perhaps that will cool you down."

Positions are drawn

The main positions in the conflict are clearly established in K.G. Ossiannilsson's letter. Folke and Karin believed Folke had been unjustly treated because he had received much less financial support than his brothers, especially the oldest brother. The father's stance was, and remained, that his income was higher when the oldest son attended the university and significantly less when Folke enrolled eight years later.

The next move was made by Folke's brother Frej in a June 1943 letter in which he expresses regret that he is unable to attend Folke's and Karin's wedding. Frej then declares: "In my capacity as the oldest brother I feel that it must be my duty to request that you immediately reconcile with your parents."[17] This injunction was not an effective strategy for negotiation, especially since Frej then continues with several pages of invective, wholeheartedly taking his father's side in the conflict. Frej's attitude also reflects the old-fashioned notion that the oldest brother retains a certain responsibility for his siblings even when all concerned are adults.

Frej's letter provides financial statistics: he had received more than twice the total amount of money given his younger brothers Göran and Sölve. Folke apparently received nothing at all. Naemi had expressed the belief that Folke should be compensated later, in the parents' will. This was a possibility K.G. rejected.

The long letter from Frej also includes a passage accusing Folke of pro-German sentiments during the war. Independent sources do not corroborate this charge. Gunnar Westin, a graduate student with Folke, remembered that Folke was generally regarded as apolitical and emphatically not as pro-Nazi. Even other fellow students I have interviewed, such as Torsten Husén, do not remember German

16 PA-MS, Karl Gustav Ossiannilsson to Folke Dovring, 10.12.1942.

17 PA-MS, Frej Ossiannilsson to Folke Dovring, 26.6.1943.

sympathies. The previously cited letters from Frej to Folke about the anti-Nazi journal and from K.G. to Karin about German bombs wiping out culture also weigh against Frej's subsequent allegation. But in family conflicts words, once spoken, are often misinterpreted.

Interestingly, in this letter Frej also distances himself from his father's earlier admiration of Hitler and Mussolini, claiming he opposed K.G. silently for more than ten years. It is difficult for a young man to pursue such a strategy, especially when important matters are at stake.

Folke however did not in this conflict, or any other, choose the strategy of silence and acceptance.

A second chance

For nearly a year there was no contact between Folke and the family. In 1944 Folke and Karin made overtures to bridge the gap. A letter from Folke to Naemi in May, on Mothers Day, was followed by a visit to Frösta, though K.G. chose to be away, thus marking his disapproval. After that, additional letters were exchanged with both parents.

In a surviving letter, Karin tells her in-laws about getting together in Lund with Folke's brothers and their children. Frej's little daughter Maj is described as "lively and flourishing".[18] Karin also expresses her thanks for the parents' help to Folke, which had meant a lot to him: "Both economically and psychologically." Apparently a rapprochement had been effected, and Folke had been partly compensated. (Different numbers are mentioned in the letters, but apparently Folke got about 4 000–5 000 kronor, a little less than the other younger brothers, and much less than the oldest.)

In July 1945 a letter from K.G. to Folke further smoothes over past disagreements.[19] The father gives a long description of a large social gathering where he had met the Crown Prince of Sweden. Ossiannilsson also discusses plans for the future: books sketched out, radio appearances scheduled. At this time Ossiannilsson was on the verge of a comeback; one of his novels was even made into a movie. His letter breathes optimism, an overlooked author's delight at being back in the spotlight. Goodwill is touchingly expressed in small details such as K.G.'s gratitude for the strawberries Folke and Karin had served him when he visited them in their new apartment.

The second challenge

The next and definitive break between Folke and his father came when Naemi was committed to a mental hospital in November 1945. Folke fought against this step

18 PA-MS, Karin Dovring to Karl Gustav and Naemi Ossiannilsson, 20.11.1944.

19 PA-MS, Karl Gustav Ossiannilsson to Folke Dovring, 18.7.1945.

and questioned the severity of her illness. He even initiated a litigation, to bring her home to Frösta again.

In January 1946, Ossiannilsson gave Folke a further 2 000 kronor as a loan and he mentions that he has to pay for his wife at the mental hospital. In the accompanying letter he also recommends that Folke postpone his planned dissertation.[20] Folke replies by requesting a complete account of all financial gifts, and he also implicitly rejects the idea of postponing the dissertation. He wants to finish his work and be first among the brothers to earn the doctorate.[21]

A year later the crisis was a fact. Folke had tried to hinder his father from being the legal guardian for Naemi. His argument was that the father had not visited his wife for a whole year.

In a letter to Folke of early January 1947, Frej declares his unqualified support for his father's action.[22] He even wants to put off visiting his mother in the hospital to avoid causing distress and states that he opposes Folke's goal of bringing Naemi home again. Frej concludes by defending the housekeeper; Vivi, he claims, is needed at Frösta to keep things running smoothly. If Naemi were to come home again she would send Vivi away, and his solution is that this must be unequivocally rejected.

Ossiannilsson at the same time sends a letter to Folke and tells him that he soon will visit Naemi. He also defends his standpoint that she is to be kept at the mental hospital. He mentions suicide attempts, and he claims that the last morning at home she said "unmentionable" things to him before she was sent away. Three days later he has visited her, and was in agreement with the head physician at the mental hospital, John Nordström, that she had to stay. He finishes off his letter with the advice that Dovring instead should think of his dissertation.[23]

The end of this conflict came when Folke received a letter from K.G. dated 23 January 1947.[24] Not a word is said about Naemi. Instead K.G. expresses thanks for the copy of Folke's dissertation that had just arrived.

The copy of the published thesis sent to K.G. Ossiannilsson has survived. Folke signed it "To Dad from Folke."[25] In the last lines of the Acknowledgements the author also expresses gratitude to his parents for all the years of home schooling. He notes as well that they were the ones who "first aroused my interest in

---

20 PA-MS, Karl Gustav Ossiannilsson to Folke Dovring, 9.1.1946.

21 PA-MS, Folke Dovring to Karl Gustav Ossiannilsson, 19.1.1946.

22 PA-MS, Frej Ossiannilsson to Folke Dovring, 11.1.1947.

23 PA-MS, Karl Gustav Ossiannilsson to Folke Dovring, 12.1.1947 and 15.1.1947.

24 PA-MS, Karl Gustav Ossiannilsson to Folke Dovring, 23.12.1947.

25 In an earlier book, Fogdö (Vårfruberga) klosters jordebok, from 1945 the dedication was: "To Mom and Dad", but in 1947 his mother did not any longer live at Frösta. Both copies are now owned by Alf Ericson, Linköping, Östergötland, my Ph.D. student doing research on "attungen", cf. footnote 80.

studying history",[26] hardly surprising in a family where the father was a historical novelist. Folke had apparently not understood how deeply he had hurt and challenged his father.

Before publishing the thesis, Folke changed his surname from Ossiannilsson to Dovring, the name of Naemi's family farm. K.G. Ossiannilsson was not entirely pleased with Folke's new surname, commenting in the letter that Folke had seldom followed his father's advice. But K.G. accepts his son's choice, recognizing the significance of its timing to coincide with the presentation of the young man's first major publication.

The father is even less gratified, however, by Folke's expression of thanks in the Acknowledgements, believing it implies that all support from his parents ceased when he enrolled at the university. This is a biased reading of the text, apparently based on years of contention and strife.

Having chosen to see his son as an ingrate, K.G. continues along the same line, itemizing all the help Folke has received over the years: clothing, food, money in the form of gifts or loans. The father had kept an account of all costs he incurred raising his sons, a total of 30 000 kronor, which today (the year 2009) would be more than 500 000 kronor – which is about 70 000 U.S. dollar.

The father then, in his long letter, accuses Folke of intellectual arrogance, of believing he is the most gifted of the brothers. This comment could illustrate an essential feature of all family conflicts, the misinterpretation of an antagonist's position, but it could also refer to an implicit competition, especially between the oldest and youngest brother, to establish who was the better scholar.

After some further comments on earlier matters of contention, K.G. brings out the heavy artillery. Folke is a person for whom others are either enemies or tools. Karin is described as the one who has fanned the flames of discord, perhaps even lit them. Her goal, according to K.G., is to split up the family because she herself came from a broken home. The patriarch declares, "If you want enduring success and joy in your life you will have to change your behavior toward other people".

As a final blow, the father demands that Folke immediately, before April 24, repay the loan of 2 000 kronor that he had received the year before. K.G. gives specific instructions about how the promissory notes should be redeemed. Karin and Folke had to borrow money to meet K.G.'s ultimatum. With Folke's rather low salary it would be years before they were out of that debt.

This letter from K.G. reveals a degree of rage not in evidence elsewhere in this long conflict. What prompted it? The name change? Certainly not. Earlier skirmishes? Probably not. The reason for his emotional upheaval is surely the controversy studiously avoided in the letter, Folke's attempt to bring his mother back to Frösta.

26 Dovring (1947), p. 6.

However Folke did not give up that easily, and the rest of 1947 he fought for his mother's right to come home. At last he received a letter from the head physician, dated November 1947, which closed the case.[27]

The doctor, Nordström, begged Folke to be realistic. This could be a public scandal, which would hurt Naemi even more than staying at the hospital. The father fought "with somewhat overstated means, admittedly", for a respite. One must try to understand him. He had to continue with his work, and he could not be alone. In his letter Nordström told Folke that the goal of Naemi, to get her husband back, would never be realized even if she were to return home to Frösta. The father threatened divorce if Naemi was sent home.

From a gender perspective this conflict is very interesting, and it was not all that unusual that a forsaken wife went mad. We have other examples, not the least of which are from intellectuals in early twentieth century, and my hypothesis is that new female ambitions were met with male counter-reactions. Naemi could however have fallen ill and have been hospitalized for other reasons. I do not intend to follow this further. It would demand a more extensive study and lead away from my topic. The reason why I treated this subject is mainly that, according to my opinion, it came to steer Folke's life.

The mother stayed hospitalized the rest of her life. There are about a hundred letters among the Dovring papers. In many of them Naemi seems to be rather confused, in others perfectly normal, talking for instance about how much her doctor Nordström likes to read what K.G. writes. Both Folke and Karin corresponded with her, but not during the years when Naemi lived with her oldest son, Frej. In the earliest letters she now and then asks Folke: "when can I come back home?"[28]

A definite break

After this battle, all contact ceased between Folke and his father and oldest brother. Many years later, in the 1970s, after K.G. Ossiannilsson had died, Frej wrote to Folke again, trying to reestablish relations, but Folke did not reply.[29]

In the unacknowledged competition between the two brothers, Folke was the first to defend his thesis, in 1947. Frej, eight years older, presented his in 1949. Folke was appointed Professor of Agricultural Economics in 1960 at the University of Illinois, while Frej got the formal name of professor in 1963 at the University of Agricultural Sciences in Uppsala, but that is not quite the same as being full professor. Folke produced many more Ph.D.s and more books, but Frej's works on insects are still quoted among entomologists the world over.

Jan Pettersson, one of Frej's two doctoral students and his successor at the

27 PA-MS, John Nordström to Folke Dovring, 17.11.1947.

28 PA-MS, Naemi Ossiannilsson to Folke Dovring.

29 PA-MS, Frej Ossiannilsson to Folke Dovring, 11.12.1978.

University of Agricultural Sciences, provides some insight into Frej's academic career and personality.[30] Frej Ossiannilsson had many international contacts, and, because he was an expert at identifying insect species, he spent much of his time responding to queries from other scholars.

He seldom spoke with anyone, instead spending most of the day peering into his microscope. If Jan had questioned his supervisor, Frej would look up from the microscope, listen politely, and then turn to his microscope again. Several hours later he might approach Jan to give his opinion with a brief justification. Frej never mentioned his family, and certainly not Folke. It was a surprise to Jan to hear, during my conversation with him, that Frej had a brother in America. This strategy of complete silence had been refined early on and was Frej's inheritance from the family conflict. Nonetheless he had a kind of humor: on the cover of his thesis, on the drumming sound of insects, there is a drawing of an insect with drums as a jazz-musician.[31]

The position of rebel and defender of justice, a stance Folke Dovring embraced on several occasions in his life, may also, in part, be traced back to the drawn-out family dispute. When Folke chose to stand up against his father, no middle way was possible, no long-lasting compromise was available. While he was engaged in this battle, he was also writing his thesis, in which he attacked leading scholars among historians.

It would nevertheless be an oversimplification to claim that Folke became a rebel solely because of the family constellation. Those who devote their lives to opposing the status quo do so for a variety of reasons, including the fact that there are, in fact, wrongs to be righted.

There is yet another fascinating aspect of the story. Folke's career reveals obvious parallels to that of his father: a remarkable beginning, with several important books published during the first ten or fifteen years, followed by a long period that produced solid, respected work not considered equally significant today. Since father and son were active in entirely different fields and embodied quite divergent political and ideological positions, this similarity should not be overemphasized. Still, it raises interesting questions. Do we tend to recapitulate our parents' behavior, even if we detest them? Have many of us in some shadow-like way repeated features in their lives and careers?

But the particularities of every life must be stressed, and in many respects Folke's career and ideology came to be vastly different than his father's.

30 Interview with Jan Pettersson, 22.3.2002.

31 This cover has got some attention from librarians, see Martling (2002).

CHAPTER 2
Medievalist (1934–1951)

Standing in a big library, for instance a major American university library, you can hear the writers trying to talk to you, and tell you the best and most intelligent ideas they have. Take a book from the shelf, catch a hint of its main thrust, and if you become interested, read further, or perhaps even borrow it for closer examination. Writers must be allowed to communicate through their books without interference from familiarity with them as individuals or awareness of how successful they eventually became.

Here I shall let Dovring speak through his books on history, especially the first two important books about Swedish Middle Ages (published in Swedish, but they include English-language summaries). But I will not let Dovring speak alone. A book is never a monologue. The statements of the writer, the subject chosen, the construction of the arguments, always arise in the context of a dialogue with other authors and scholars. This is the kind of discussion one experiences in a major library. Scores of voices speak to each other, each trying to be heard above the rest, or at least together with the others.

In the first lines of his 1947 dissertation Dovring declares himself an agrarian historian (Swedish "agrarhistoriker"). He continues, "An agrarian historian does not look for isolated facts, but for typical facts".[32] This sentence characterizes his endeavors in the years to follow, years that brought him to the forefront of European research but at the same time alienated him from Swedish historians.

He was not the first in Sweden to call himself an agrarian historian, instead time seems to be ripe for this sub-discipline of history in Sweden as in other European countries. One of his fellow Ph.D. students wrote about crofters, and labeled himself "agrarian historian" in his dissertation from 1945.[33] In the years 1943–1948 the first two volumes of "The history of farmers in Sweden" were published.[34] During these years even a professorship in agrarian history situated at the Swedish College of Agriculture (Lantbrukshögskolan) was discussed, but

---

32 Dovring (1947), p. 7.

33 Elgeskog (1945).

34 Ingers (1943–48). Enoch Ingers died shortly afterwards publishing the second volume and the last, and third, volume did not appear until 1956, written by Sten Carlsson. Ingers express his thanks in the Acknowledgment to both Dovring and Elgeskog. Later Elgeskog in his Acknowledgment thanks Dovring for proof reading. Apparently there was a small group in Lund around 1945–1949 that considered themselves as agrarian historians and helped each other, see Myrdal 2008.

was turned down by the staff and the students.[35]

What Dovring did was to raise the temperature in agrarian history and in medieval history, by questioning those who looked for “isolated facts”? Without a doubt this was an attack on most of his fellow historians in Sweden, who concentrated their research on individuals of the nobility.

European historiography

From late nineteenth century, a trajectory in the development of history as a discipline was the division between those who describe the upper strata of society and those who want to emphasize the history of the common people.[36] When democracy pressed itself forward in the European societies, an ideological consequence was that the historical importance of ordinary people was recognized. But it was not the historians themselves that stood in the forefront of this change. History as a discipline had its roots in chronicles written by, for, and about the upper classes. Professors in history generally stood to the right and resisted the new approach. Step by step the balance shifted in favor of history of the masses. It was a process marked by battles, as the famous Lamprecht-controversy in Germany around 1900. Karl Lamprecht advocated a mainly non-political history, which he categorized as “cultural history”. Nearly all of the professional German historians turned against Lamprecht, and cultural history was for a long time looked down upon in Germany.

Another controversy in late nineteenth century concerned methodological questions. As history gradually became the subject of scientific scrutiny, historians were forced to confront the fact that every historical document is biased in one way or another. This was a traumatic realization that caused many established truths to be questioned, undermined, or rejected. Usually Leopold Ranke in Germany is identified as the founder, in the first half of the nineteenth century, of this new and more critical approach, but many historians were already moving in the same direction. History changed from an art to a craft, where skill at interpreting the sources became the sign of craftsmanship.

This new historical methodology did not at first encourage research on the

---

35 Personal communication with Bengt M.P. Larsson (who died unexpectedly in 2008). He did research on the early introduction of agrarian history, and also presented his results on a seminar 2008. His main results will hopefully be published soon. He proved that the scientist had a rather narrow perspective, and therefore students in agricultural economics became afraid that resources for social sciences would be split if agrarian history was to be established at the College, and thus opposed agrarian history as a subject. Outside the college, agrarian history had support from politicians and some historians during the years 1948–50, see Odén (1991), p. 145, 165.

36 About general historiography see for instance Breisach (1983), Burke (1992), Raphael (2003) and Iggers (2005).

masses. The sources scrutinized in new ways were the same ones historians had been working with before: those that described kings, the nobility, battles, statesmanship, and diplomacy. Political history, and especially the history of the state, remained the focus of interest for the source-critical school of historians. As Peter Burke has shown, Ranke's transformation even tended to marginalize broader perspectives such as social or cultural history.[37]

A methodological change in congruence with research about common people came with the growing importance of quantitative history. This started early on in Economic History, but was not brought to bear on other sub-disciplines until after World War II. Then the shift was so profound that quantitative history for a time became dominant in most historical research, especially in social history.

The new paradigm after World War II often took the form of an adherence to the French *Annales* school. In the end of 1920s some young French historians, under the leadership of Marc Bloch and Lucien Febvre, started a journal named *Annales d'historie, économique et sociale*. They took up broader perspectives in history, mainly in opposition to traditional history. Their aims were to write a problem-orientated history instead of narrative history, to take up a whole range of human activities instead of political events, and to do interdisciplinary work.[38]

After the war the historical community in the world had to be reconciled. This work started immediately, and politically radical historians such as the Norwegian Halvdan Koht had prominent roles. Koht had been one of the most prominent inter-war historians in Europe, together with for instance Henri Pirenne. Koht served as foreign minister of the Norwegian exile-government in London, and went to the U.S. during the war. He was, in himself, the continuity of the European historical society. The first general international congress for historians after the war was to be held in Paris in 1950. After some debate it was decided that even the Germans should be invited, to mark the new concord of historians.

The organization of the congress was handed over to a group of young *Annales*-historians, such as Charles Morazé. They worked out the program for the congress in collaboration with well-known historians such as Lucien Febvre and Fernand Braudel. The *Annales*-historians monopolized or at least dominated the congress; they set the agenda for discussions. This was the start of a more widespread acceptance of their ideas.

Especially in the field of economic and social history, the discussions were extensive. For instance the English historian Michael Postan, affiliated with the *Annales*-historians, talked about the long waves in medieval history. He was attacked by more traditionalistic historians on the basis that the personality played a larger role during the Middle Ages than in other periods. But the traditional historians were on the defensive. The congress marked, not least regarding the Middle Ages, the

37 Burke (1990), p. 7; Burke (1992), p. 5–7; Iggers (2005), p. 28.

38 Burke (1990) passim.

start for social history as the dominant paradigm in historical science.[39]

During the following international congresses, first in Rome in 1955 and then in Stockholm in 1960, this debate moved more into the background as the breakthrough for these ideas had already occurred. Other controversies came to dominate these congresses, not the least of which was a fight between East and West over Marxism and history.

Swedish historians

In Scandinavia the fight over cultural history came at about the same time but with a partly more radical content than in Germany. A first wave of industrialization and the attendant growth in the political and economic importance of the middle class brought about an ideological change. Young writers were motivated to address political and social issues; a new market for realistic prose and drama evolved. Within the discipline of history, a similar movement was initiated. One of the most outspoken of the new historians was August Strindberg, best known today as an innovative and controversial dramatist. In 1881–1882 he published a People's history of Sweden, *Svenska folkets historia.* He declared that his work would not focus on kings and their followers but on farmers, workers and the details of everyday life. His challenge was met with an outcry from other historians, leading to "the great battle about cultural history" of 1881–1882, waged in journals and newspapers.[40] Strindberg's radicalism turned his fellow cultural historians against him, and cultural history went underground.

At the beginning of the twentieth century, however, newly established fields such as economic history and folk-life studies (ethnology) took up research about the common people as well as historical geography.

The more critical approach in reading the sources came about in Sweden at the same time as in the rest of Europe. In a second wave of source-critical studies, in the beginning of the twentieth century, a group of Swedish historians, which later came to dominate the society of Swedish historians, tended to be more severe in their criticism than most historians in the rest of Europe.[41] I will later examine their ideas. In Sweden and in the rest of Europe, this closer reading of

39 Erdmann (1987), english translation Erdmann (2005), has described the international historical congresses in detail. Erdmann (2005), p. 207 talks of a "a paradigm shift from historicism to historical social science", but also emphasizes that social history took a large part of even earlier congresses.

40 About this conflict see: Myrdal (1982).

41 In international historiography this Swedish group of historians is seldom mentioned, but when mentioned it is their stern criticism of sources, which is emphasized, see Raphael (2003), p. 67 who talks about their "rigorosen Kritik", rigorous criticism. This kind of scrutinizing can be considered as close to positivism.

sources, especially medieval sources, tended to concentrate research heavily on the state and the elite.

This specific group named themselves "the source-critical school". The leaders were Lauritz Weibull, his younger brother Curt Weibull and, among the most gifted of the young, was Erik Lönnroth, favorite disciple of Curt Weibull.[42] They were political liberals. After fierce battles in journals and over appointments, by the 1950s, the new group had seized most of the chairs in history. A not insignificant fact is that many of the conservative older-generation historians had been adherents of, or at least friendly toward, Nazi Germany, which definitely weakened their cause after 1942.

With one important exception, interest in international history was limited. The Swedish source-critical school also had little influence on the European debate during the period from 1920 to 1950. The bitter internal fight seemed to turn Swedish historians inward. This was the intellectual environment in which, as an undergraduate student, Folke Dovring began studying in Lund. There he attended seminars of the leading professors in the source-critical school.

Agrarian history

The branch of history, which copes with farmers and agriculture and their environment, has its own historiography running parallel with the overall discipline.[43] Not surprisingly, agricultural and agrarian history was established early on as a separate entity in the United States, which traditionally defined itself as an agrarian nation. The oldest journal devoted to the subject, *Agricultural History*, founded in 1927, is likewise American.

In Europe it was in France that rural and agricultural history developed as a discipline. The Frenchman Marc Bloch published a history of rural France in 1931, but not until after World War II did the book establish an international reputation warranting inclusion in the canon of historical writing. Bloch nevertheless had many international contacts especially in England. He died at the end of the war, fighting for France.

English scholars also made some contributions to agricultural history. Indeed it was under English editorship that the first overview of the subject was published: the first volume of *The Cambridge Economic History of Europe*, which had the subtitle *The Agrarian Life of the Middle Ages*.

---

42 According to Birgitta Odén, interview 09.05.2003, Lauritz Weibull was not as impressed by Lönnroth as many others, but I do not here have the intention to write the history of the Weibull group.

43 For an overview see Myrdal (2001). During the last decades agrarian and rural history has been an active field of research, cf. all the new literature mention in books on the historiography of several European countries (England, France, Germany, Belgium, the Netherlands, Spain, Italy, Poland): Thoen & Van Molle (2006); Alfonso (2007).

The English historian Eileen Power had planned this remarkable project and especially this specific book since 1932. The group who planned the project was especially keen on getting anti-Nazi writers. For instance the editors thought Scandinavians to be honest, which meant that they were mainly anti-Nazi.[44] The editors were allied with the *Annales*-historian, and they declared that in their choice of authors Marc Bloch's "knowledge of European scholarship and scholars was always at our command."[45]

Though Power died suddenly in July 1940, the volume appeared in 1941. Her co-editor, J.H. Clapham, the leading economic historian in England, signed his preface at Christmas 1940, explaining that completing the task had been extremely difficult. This is one of the most strange prefaces ever written to a book in history. Not only had Power passed away, but authors were lost in the war or there were no later news of them. In other cases the editor was misinformed, such as in believing Marc Bloch to be "safe in America" when Bloch instead had joined the Resistance and eventually was shot by the Germans. (In the second print of the book, from 1942, the foreword reported that Bloch is safe in France.) Clapham declared that without a new co-editor, Michael Postan, the book would not have been finished.[46] Postan was a refugee of pre-revolutionary Russia, but quite influenced by Marxist ideas. He had married Power and later became the leading economic and agrarian historian in England. Postan was a medievalist, and would later approach Dovring, which I will come back to.

This book pointed to the future, not only because the editors forged ahead with publication during the war and tried to keep the international contacts open, but also because, subsequently, the history of agriculture took a great leap forward. University departments of Agrarian History (or Rural History) were established in several European countries in the 1950s; national journals were started in England, the Netherlands, Germany, Denmark, Italy and France between 1953 and 1961. This was a part of the paradigm-shift in history that came after 1950.

Parallel with this advance for agricultural history, other disciplines likewise began taking an interest in peasants. When Dovring declared himself as an agrarian historian in 1947 he thus stood at the forefront of international research. But how could a young student in a remote and quite conservative part of Europe establish himself in this position?

44 Berg (1996), p. 215. Maxine Berg has given an extensive description of Eileen Power's importance as historian in her biography, and she also presents her husband, M.M. Postan, who came to play a decisive role for the new medieval history in England and Europe.

45 Clapham (1941 I:1), p. viii.

46 Clapham (1941 I:1), p. vii, viii. For background to the intellectual milieu and the planning process, see Berg (1996).

Contact with international research

In his doctoral thesis Dovring quotes only one work published in English; the first volume of *The Cambridge Economic History of Europe,* and especially the chapter on Scandinavian agriculture by his supervisor, Sture Bolin. He was one of the leaders in the source-critical school and pupil of Lauritz Weibull. Bolin presented his thesis in 1927 in Lund, and some years later was appointed to the second chair beside Lauritz Weibull. Bolin held this chair from 1938 to 1963.

Bolin contributed to the international debate more than any other historian in Sweden at this time.[47] He had challenged Henri Pirenne's theories about a rupture in trade caused by the wars in the Mediterranean with the advance of the Muslims, a theory Pirenne presented in the 1920s. Bolin asserted instead, in articles published from the 1930s, that trade continued. It merely took another direction when a new trade route was established from the Arabic world through Russia over Scandinavia and to Western Europe.

The founders of the *Annales* journal looked upon Pirenne as one of their ideological fathers, and Bolin must have been well known to them. He was conservative, but also an outspoken anti-Nazi. Thus, when the *Annales*-affiliated editors for *The Cambridge Economic History of Europe* started to look for a Scandinavian contributor, it is not surprising that Sture Bolin was asked to contribute the chapter about Scandinavian agriculture.

Bolin had worked primarily with numismatics and trade, but he had also written about subjects related to agriculture, such as medieval demography and social organization. In his chapter he presented a survey of research about subjects such as the forming of villages, animal husbandry, crafts, class divisions, taxes and the late medieval crisis.

Contacts between the volume editors and Bolin were cut off during the war; they had not received his bibliography when the book went to press. As the editors did "not wish to hold the volume up until peace comes",[48] they decided to publish his article without it. (Bolin's article also contains an amusing misprint, as he is identified as "Store Bolin", Swedish for "The Big Bolin" – obviously he had not been able to proof read this text.)

In the foreword to his 1947 doctoral thesis, Folke Dovring noted that it was his supervisor, Bolin, who first suggested the topic: land area measurements. Dovring also mentions Bolin's knowledge of the Middle Ages and agrarian history.[49] Strangely enough, Dovring did not cite much international scholarship in his early work. Possible reasons for this omission are his assessment that it was irrel-

47 Odén (1975), p. 271. The economic-historian Eli Heckscher was even more involved in the international debate and is probably the most internationally well known Swedish historian.

48 Clapham & Power (1941), p. 610.

49 Dovring (1947), p. 5.

evant to his specific investigation and that he lacked international contacts, which made references to the international research frontier less self-evident. He probably also considered that a general allusion to the Cambridge volume, without discussing or presenting the content, was enough.

Connections to European research are nevertheless apparent. Agrarian history stood on the eve of what may be called a breakthrough, and the scholars responsible had already been in contact with Bolin before the war. During Bolin's work with the survey chapter, he asked a young graduate student in the department to research further some of the questions he had touched upon.

The subject he proposed to Dovring was related to his own discussion about villages, taxes and land ownership. And thus Bolin's chapter came to be the only Dovring later quoted in his thesis from international research.

The land area measure as a question

Bolin's chapter in the first volume of *The Cambridge Economic History* had village organization as one of its main themes. His hypothesis was that "village measures", namely land area measures, were of "great importance for the activity of the village community".[50] The organization of the village community was regulated by provincial laws from the thirteenth century and also by the first national codification of law that came in the middle of the fourteenth century.

As farmers cooperated more extensively, legal provisions established duties and rights. The layout of the village, with land area measures, was also connected with the distribution of taxes imposed on Swedish farmers during the same period (the second half of the thirteenth century). It was this process of change that Bolin asked his new graduate student to describe and analyze.

This was not an unimportant question. In the restructuring of society in Northern Europe in the High Middle Ages, changed property rights to land was an important element. Land could for the first time be bought and sold, and landowners could acquire and manage farms spread over several regions. Land eventually became the basis for taxpaying and for village cooperation. One of the prerequisites for a better-defined land ownership was measurement. Without a precise instrument to measure land it would not be possible to claim rights to a specific piece of land.

After finishing his bachelor's degree in 1939, Dovring almost immediately began to work on his thesis about land measurement systems in Sweden. Two different measures were used, the "attung" and the "markland". Attung was in use from the time of the earliest written sources, so the age of the term is uncertain. The old Swedish word may be translated "an eighth part", often, but not always, understood as one part of a village with eight farms. Markland is mentioned more frequently from about 1250–1270 and probably does not significantly predate

50 Bolin (1941), Bolin (1966), p. 644. The chapter was republished 1966 without change.

this period. There is some contested evidence of earlier use of "mark" in connection with land in Sweden, but these instances have been proved to refer to rents and not to a specific land measure unit. Markland proper may be translated "land area measured as one mark". Use of the term reflects the growing importance of money with the minting of national coinage and a growing market for trade. As such one "mark" was the basic monetary unit. The measure unit "mark" was however widely used in Western Europe, including for land measures.

In certain areas the old land area measurement attung survived, especially in Östergötland, where Folke Dovring grew up. In the political center of the country markland came to dominate, especially around Lake Mälaren. In still other areas, especially in western Sweden, no fixed system of land measurement developed.

His first task was to narrow down the period when markland and perhaps also attung was introduced. He also must more specifically try to identify why these land measures were introduced. And if he could, try to identify the actual area they compounded, which would open up the discussion about farm size, village size, etc.

His supervisor perhaps believed that Folke Dovring would confirm, build on and further prove the results he, Bolin, had presented, but that was not in Folke's mind.

Developing methods

Seven years elapsed before Dovring completed his thesis. This prolonged gestation period was caused in part by lack of funding, which necessitated part-time work as a librarian. Dovring was also conscripted into military service for several periods in the early 1940s.

Another reason for the delay was that Dovring never took a short cut toward his goal. His writing style reflects this thoroughness, and he treats each subject exhaustively before he proceeds to the next. Passages that to an uninitiated reader seem to be lengthy digressions Dovring viewed as indispensable investigations, necessary stepping stones.

Dovring's first problem was to locate and use sources that would offer a firm factual basis for his arguments and enable him to avoid sheer speculation. Preserved, written sources from medieval Sweden are rare, mostly consisting of charters. A few longer cadastres and account-books are preserved, mostly from the Late Middle Ages. However, during his work with the dissertation, Dovring in fact discovered the oldest cadastres preserved in Sweden, from late twelfth century. This had been stored in the university library in Uppsala, without much notice. By carefully editing the whole manuscript and identifying villages and farms mentioned he was able to date the volume. He also took time to publish this edition separately, in 1945, before the dissertation was finished.

Sources from the Middle Ages are scarce in Sweden, compared with, for instance, France or England. Beginning in the early sixteenth century, the source material expands considerably. The state apparatus in Sweden developed into one of the most efficient in Europe, not only producing but also preserving an enormous number of written documents. Individual farms were registered every year. From the period 1520–1620 two million pages of accounts have been preserved, and thereafter the amount of preserved material continues to expand. In the seventeenth century, the state initiated an enormous project to produce village maps, with the aim of covering every single farm and field. The project continued for decades; between 20–40 % of Swedish villages were mapped in about 50 000 village maps from circa 1630–1720.

Dovring decided to use these extensive later sources as a fixed point, from which he could interpret older documents. This was a new idea in Swedish historical research. Earlier medievalists had not used these later sources but concentrated on a restricted number of early documents mentioning the attung and the markland. When they examined these documents over and over again and tried to squeeze information from them this led to a flood of speculations, many of which were precarious because they were based on only two or three documents. A stalemate was reached which Dovring intended to break.

Dovring built a collection of examples of farms and villages with sources to obtain a sound base for a statistical interpretation. The examples should be of "high quality", i.e. they should give extensive information about the questions Dovring was interested in, especially how much land area one attung or one markland represented. Dovring collected over 200 examples in his database. Large numbers of cases had been tested and turned down as not sufficiently informative. About 40 % of the pages, printed with small-size type, in Dovring's dissertation consist of such data – this is more than half of the text. First presented are the medieval documents, then evidence from the sixteenth century accounts, and last village maps.

This quantitative and retrospective approach was in line with the international trend. Indeed Dovring's approach was later to become the model for all studies about the economic and social history of the Late Middle Ages in Sweden. He called this database a *kasuistik* (casuistry), a term originally used in philosophy, law, religion and later also medicine meaning a collection of examples. Since Dovring assigned the term a specific meaning, I will use the Swedish term. Dovring was the first to apply this term to historical research. Some years after Dovring's book was published *kasuistik* became a common term among Swedish medievalists. They made databases of farms with medieval and later evidences combined. The *kasuistik* was expanded to register all farms in a region, not just a selection of those with most data.

One of my main oral sources is Gunnar Westin, who was a fellow graduate student at the department of history. He remembers how happy Dovring was when he had found out that he could borrow this term from law.[51]

For certain reasons Dovring's pioneering contribution to the methodological development was not acknowledged by later research in Sweden. Even among medievalists, who themselves were early adopters to use the term *kasuistik*, there is an uncertainty about from where this term originally came. This is supported not only by personal comments, but also by the fact that in dissertations and books from 1960s in Sweden, where the term was used extensively, no one referred to Dovring as the scholar who coined the concept for this purpose. This is however not the whole truth. Dovring's database came to be known as one of the most reliable, and is today regularly used for all types of research about medieval economy and settlement.

Medieval division of the land in Europe

The general European pattern was rather well known when Dovring wrote his thesis; Marc Bloch had described it in his history of *French Rural History*, and again in the first volume of *The Cambridge Economic History of Europe.* Dovring had read this article prior to writing his dissertation, as he referred to the whole book in his reference list. Later Dovring's results would lead him to examine the European connection, and point at similarities between Northern Europe and Sweden.

A long term trend through the Early and High Middle Ages was the break up of large units. From a modern point of view, it can be described as a dividing up of large land area units. From the contemporary point of view, it was probably more seen as a splitting up of large social groups, which can be labeled as the change from a "tribal" to a "feudal" society. In this gradual dissolving of large units, the "mansus" came to be the basic unit of assessment in the Early Middle Ages.[52] Marc Bloch wrote about the mansus: "There is no more mysterious institution in all agrarian history."[53] The enigma was the European extension of an institution where no common origin can be identified. In Romance lands it was called "mansus", in England "hide", in Germany "Hufe", in Denmark "bol".

From the ninth century the mansus was used for rough estimates of acreage or value of the land, and used by the king as a basis for various taxes, utilizing a system already established. The mansus was kept as a unit according to the standard within a region, and these units were then further divided into fractions of one-half, one quarter and even one-eighth. The process was combined with the development of a more precise land area measurement. Even smaller units so land could be transferred. Land measurement would have been a strange idea in a so-

51 Interview with Gunnar Westin, 3.2.2002.

52 Bloch (1966), p. 155–161.

53 Bloch (1966), p. 277. The chapter was republished in the second edition of *The Cambridge Economic History of Europe* in 1966 without change.

ciety largely based on families and kinfolk, but it now became a part of a whole new social web that emerged between 800 and 1200 in Northern Europe.

Later research has mainly confirmed Bloch's description, even though there has been discussion about the origin of the institution, the importance of slaves, and also of how the estates were formed "bipartial", with a demesne cultivated directly for the lord, and other holdings worked by tenants.[54] I do not however here intend to dive into the vast discussion of feudalization and social change around 1000 AD, which is a major topic in contemporary historiography of the Middle Ages.

Supposed Swedish particularity

A focal point in Swedish historiography has been to what extent the country has had a separate development from the rest of Europe. The standard interpretation was that this periphery had a history of its own. A free peasantry from time immemorial constituted an exception from harsh European feudalism. This particularity became an important part of the Swedish self-esteem – not the least in connection with the social reforms of the early mid twentieth century, which made Sweden an example to the world. Dovring gradually developed a different position from the main stream in Sweden concerning the Swedish particularity.

Conventional Swedish historians could not accept an early hierarchical society, and thus tried to find the origins of assessing land in early village organization: free peasants collaborating to structure the society. In Sweden, when we arrive at the time when the state was formed, a real kingdom, we are in the High Middle Ages (circa 1000–1200). This process was seen as a crucial change of the whole political and social structure; when the free peasantry to a certain extent became subjugated. However for a long time no general theory was advanced. The Swedish historians, being very empirical, hesitated to pursue such an endeavor.

Some years before Dovring started his career, the young historian Erik Lönnroth had forged a remarkable career, writing his thesis 1934 at the age of 24 years, and attaining full professorship at the age of 32. He was the crown prince in the source-critical school, and he was the favorite pupil of Curt Weibull, the younger of the Weibull-brothers. Many considered him a genius.

Lönnroth worked broadly involving large areas of study of the Middle Ages, but always with the state formation and medieval politics as his focus. He introduced the economic base as being one decisive factor. This had an enormous impact on his contemporaries, particularly the newer researchers. They had not previously considered this aspect, or even thought along these lines before.

---

54 Important contributions are for instance Duby (1968), Poly & Bournazel (1991) and Verhulst (2002). On Dovring's research on the European mansus see below.

Lönnroth took up the importance of the land area measurement and proposed a beautiful model where a system shaped from below, in the village, later was turned against the farmers in the process of their subjection. The attung was developed as a way to regulate the village community. In the middle of the thirteenth century a new system of taxes was laid out. At that point the markland was introduced as a way to regulate this new tax system. One of Lönnroth's contributions which has stood the test of time was that he firmly dated the introduction of the markland as a land area measurement to the late thirteenth century.[55]

Dovring's supervisor, Bolin, relied heavily on Lönnroth's ideas when he wrote his article in the *Cambridge Economic History of Europe*. Indeed Lönnroth quoted earlier works by Bolin and other colleagues. A new consensus took form about the Swedish particularity.

Dovring's theory about the genesis of land area measures

Dovring in this book systematizes the main factors behind the introduction of land area measures in: measure of land in relation to the villagers to allocate obligations to the village; measure of land to regulate the amount of rent paid; measure of land to allocate taxes. All these factors had a role in forming the land measure, but the problem was which of them had been most important for the introduction of the land measures. During his work with the dissertation he came, on empirical grounds, to reject nearly every part of Lönnroth's model, though he accepted, and even further confirmed with supplementary evidence, Lönnroth's re-dating of the markland to the late thirteenth century.

According to Dovring, the attung-measure originated as a regulation of the oldest duty to the state, the duty to take part in the naval warfare, ledungen. It was an imitation of a Danish land-measure system introduced during the Viking Age, which also was connected with the ledung. The attung denoted parts of whole village-units, called "bol" that had to furnish men and equipment for the fleet. The system was old and difficult to assign a date, but it was probably introduced around 1000.[56] Later in the thirteenth century the attung had turned into a measure for rent and also for the village communities' distribution of rights and duties on a local level among individual farms.

Dovring thought that attung had been more widely distributed, and then replaced by the markland. He emphasized the relationship between the two systems, the attung and the markland, as an argument for a continuum, with a basic

55 I have discussed Lönnroth's, Dovring's and other scholar's contributions in Myrdal (2008b).

56 Modern research has been able to date the introduction of the attung to late 11th century, see Ericsson (2007), who by reevaluating documents earlier overlooked thus can confirm Dovring's assumption.

relation two attungar = one markland. The relation between an attung and a markland developed more gradually, and became more fixed only during the fourteenth century, with a growing land market.[57] Indeed these land measures were not a result of a policy decided from above, but rather from a process driven by economies at the local level – in combination with an overreaching change of the social system.

A direct attack on the Lönnroth's model was that Dovring was able to show that the state normally, at this early date, took out taxes on a collective basis. The early state had no way to enforce a more detailed land area measure system. This argument sweeps away Lönnroth's connection between the markland and the new taxes – both introduced during the same decades in the late thirteenth century.

Instead Dovring pointed to another part of the new socio-economic system evolving in Sweden at this time. Private landownership was introduced, and a system with tenants paying rents was established. The land measures were a way to control the amount of rent, especially as a fixed relation between one unit of land and one unit of rent in grain and money was established.

One of the main advantages of the new markland measure was that it made it possible to talk about smaller units of land. In the attung system it was not possible to measure units of land smaller than parts of an attung, but in the new system one could measure units of land down to the penny-land (in Swedish penning), which was a 192nd-part of a markland. According to Dovring, the peasants had no use for such a detailed system. It must have been the wish of large landowners to regulate the rent for their tenants and to control the amount of land they owned that was the driving force behind the new land measure system. The Church, with its organization and European contacts, also could have played a decisive role in forming the new system.

When Dovring later continued to do research, he gradually came to realize more clearly that the Swedish system was a part of the European. He then also came to oppose the idea of a primordial society of free peasants.

Dovring not only questioned Lönnroth's model, but also considered Lönnroth as being careless and pointed out that Lönnroth had made miscalculations: not taken all sources into account, not understood the details of the monetary system, not understood the structure of the cadastres, etc. He felt that he, understanding the sources better and with the database he had built, had a more sound basis for assumptions. Dovring was mainly correct in his critique of details, but, to Dovring's surprise, Lönnroth was forgiven for his mistakes, as his talent for model-building impressed the academic society. Dovring would throughout his whole life refer with disdain, to intuitive scholars admired by others, thereby referring to Lönnroth.

57 Ericsson (2007) has reworked the whole source-material and proved a somewhat more complicated numerical relation between the "attung" and the "markland", even if Dovring was basically correct.

Today Lönnroth's theory about the government pressing a new land area measure system onto the population already in the middle of the thirteenth century has been rejected. Dovring's explanation has been more accepted, but generally mono-causal explanations are rejected, and scholars today prefer a multi-causal explanation where the origin of the land area measure is of less interest than the gradual expansion of the tasks linked to the societal instrument of defined land area units.[58]

Widening the question to taxes

After the dissertation he went back to his work as a librarian. However, after a while he obtained a position as assistant professor and was able to continue with full-time research. In a letter from late 1947 to his mother Folke Dovring wrote that he planned two projects: one about King Gustav Vasa's reforms in the tax system, and one about the post-medieval laws in Sweden.[59]

The latter-mentioned project was never realized, but Dovring developed an interest in the history of law. One of his idols, much admired by his mother, was K.G. Westman, who had been both a professor of history and history of law in Uppsala. While Folke waited for an appointment as assistant professor, he took classes at the law school at the University of Lund.

His interest in the history of law was clearly expressed again when he applied for a chair in history in 1951. He then outlined a project where he wanted to investigate the provincial laws from around 1300 to see if an older layer of regulations could be unveiled. This is a notoriously difficult problem, and most modern scholars have avoided it. Dovring had some ideas that could be worth testing, but he never continued with this project.[60]

He started with the first-mentioned project, about the tax-system from the Middle Ages to the sixteenth century. This was partly a follow-up of his earlier book. Dovring's method was similar to the method he earlier used for studying land units. He drew up catalogues of taxes in every district. There were over 60 different districts in Sweden, and in every district there were tens of taxes. Dovring registered hundreds of different taxes. Again the catalogue covers more than half of the text in the book.

His goal was to reconstruct the structure of the system and how it evolved during different periods. But studying taxes in historical times is a complicated subject. Tax systems were never (and are still not) constructed to be simple, as they are the result of negotiations. Furthermore, regulations seldom tell the whole truth.

58 See articles in Ericsson (2008).

59 PA-MS, Folke Dovring to Naemi Ossiannilsson, 1.12.1947.

60 He published some small articles, for instance Dovring (1947b). In the rough project draft he wanted to take up details as coins and fines, to establish an inner chronology.

Fraud from below and blackmail from the top lead to a different reality.

The method is retrospective. The registration of the tax-system in the early sixteenth century is the basis for his catalogue. From this he goes back in history to try to establish when taxes changed or were established. He stressed the need to go to the medieval sources; later sources can only be the starting-point.[61]

The reason why the sixteenth century could be this starting-point is that during the time of Gustav Vasa (king 1523–1560) the different medieval taxes were registered. Thereafter the taxes were adjusted and raised.[62] Standardization by a strong state, and at the same time an increase of the tax-level is not unusual. This was enforced in Sweden in the sixteenth century with a high level of efficiency, which included a thorough registration of the diversified Late Medieval tax-system.

Originally, in the thirteenth century, regular taxes were derived from the duty to take part in naval warfare and from the king's right to demand food for himself and his men during his travels around the country. But Dovring could also identify a number of other small taxes in a complicated system.[63]

A new phase came in the second half of the fourteenth century. After the Black Death the political organization and the social structure began to fall apart. Several additional land taxes were levied on the peasants. Around 1400 order was restored to the northern countries and for a time Denmark, Norway and Sweden formed a union with the first-mentioned country in a leading position. New additional taxes were lumped together forming a new annual tax. It was heavy but not as largely unregulated as during the foregoing period of raging civil wars.

In the early fifteenth century peasants protested against the heavy tax burden and eventually the taxes were reduced. The end of the Middle Ages was a period characterized by peasant revolts, or rather conflicts, where the peasant and their armies played a significant role in a series of civil wars. These eventually lead to the break up of the union. As a consequence of ongoing civil wars several districts were able to negotiate their own taxes, and the tax system became a mosaic of regional taxes. Nearly every district (*härad*) had its own system. This was what Gustav Vasa, king of Sweden as a separate country, wanted to register and reform.

The last phase Dovring studied came in the sixteenth century. Central power grew stronger, and a new and uniform tax system was introduced all over the country. Dovring could definitely prove that, for the first time, the state taxes were actually connected with the land area measurement system, the markland.[64] The tax became, like the rent, fixed in relation to the land unit; eventually land

61 Dovring (1951), p. 12.

62 Dovring (1951), p. 7, here and many other times he quotes his predecessor in this field of research, Hans Forssell.

63 Dovring (1951), p. 13–48.

64 Dovring (1951), p. 120.

rent and state taxes were combined to one system.

The Medieval models for taxpaying were: district assessment, assessment on persons or groups of persons and estate assessment in relation to land and in relation to cattle. Dovring showed that collective taxes were abandoned in the middle of the sixteenth century, and an individualized tax system was introduced. Dovring's description of this whole process from the thirteenth to the sixteenth century still hold as one of the best and most firmly based analyses. It is regularly quoted.[65]

He also discussed other topics. Quite a large part of the book is devoted to tax evasion. What we see in the data is mainly what the taxpayers ought to pay, but he probably overestimated the elliptic element of the taxes. A decrease in the tax-level in the Late Middle Ages was not, as Dovring suggested, to a high degree counterbalanced by the introduction of small and hidden duties.[66]

His database can be, and has been, used for other investigations. A large part of the tax was paid in kind. The nature of the products used often reflected a surplus in that specific region. Along the coasts fish was paid as tax. In the woodlands in Southern Sweden, the farmers paid with ham from acorn-fed pigs and with honey.[67]

In this book a more mature and less polemic scholar appears. He is much more conciliatory toward Erik Lönnroth, naming a couple of areas where he agrees with Lönnroth. But Dovring could not abstain from pointing out and questioning what he considered as speculations without any foundation in the sources put forward by Lönnroth.

Dovring was now developing into a leading medievalist. As both taxes and land area measures are essential for understanding the medieval society, he seemed to have made an important contribution to Swedish research in history.

Extension to Europe

As a former librarian, Dovring had been interested in the library owned by the Swedish queen Christina, who in the seventeenth century abdicated, converted to Catholicism and went to live in Rome. As a queen, she wanted to shape a center of culture in the Swedish capital, and tried to attract leading personalities in the field of science and culture. Several came, such as the philosopher René Descartes from France (who died in Stockholm), and the well-known scholar and lawyer Hugo Grotius from the Netherlands.

Dovring found that Grotius had left large parts of his library in Sweden, when

65 See for instance Österberg (1970).

66 Norborg (1958).

67 Later research often refers to his results. One example regarding medieval beekeeping has been able to build on his work, and correct him in details, see Husberg (1994), p. 104, 112, 131.

Doctor Folke Dovring while an associate professor in 1949.

Discussing an old book, probably by Hugo de Groot, with Karin, 1949.

he served the Swedish crown, and Dovring actually discovered formerly unknown manuscripts and notations by Grotius. He then became engaged in a project editing the text. He worked with this project from 1949; the book coming out in 1952 with separate articles published 1951 and 1953. The edition was widely reviewed especially in the Netherlands but also in other countries. Historians of law were impressed by this discovery, but perhaps a little disappointed that the notations did not give more of interest.[68]

Besides writing about Swedish taxes, Dovring also started a European investigation, following up his research about land area measures. It all started when his wife Karin got a grant to study archives in Switzerland for her research. Folke followed her on this journey and worked in the archives. As most of the charters and cadastres were in Latin he could easily read them.

In the following years he published quite extensively about these investigations, nearly a hundred pages spread into four articles in French in different journals. Three of them were detailed village-studies from Vaude (Switzerland) 1950, Normandy 1952 and Alsace 1952. One of the articles was a general presentation of the method, published 1951. Dovring several times points to the need for a larger European project using his methods, and probably he saw his articles as the start of such a project.[69] The scientific question he took up was the mansus, and the dissolving of the mansus during the High Middle Ages. Dovring mainly worked with the sources, and did not present the state of research. He quoted Marc Bloch and several times he referred to another of the French agricultural historians, André Deléage, for further presentation about the foregoing discussion.[70]

Dovring emphasized the village study as the only way to avoid speculations. He explained his "casuistry" method. A specific case is chosen. One must always start with a map that is as old as possible. In most cases, one could find maps of villages and their fields from the eighteenth century. From this map one must try to go backwards. As cadastres and registers often are quite conservative in the naming of fields in the same order, younger registers are modified transcripts of older ones, and the single fields can often be identified – if the sources

68 Most of the reviews were in Dutch, but in English Robert Warden published two short notices in *The Law Quarterly Review* 1949 and 1953.

69 When he did research in Rouen, Normandy in the summer 1950 he was actually interviewed by the local newspaper, and he mentions his parallel investigations in Alsace and Switzerland.

70 Dovring (1952b), p. 53 and also in other articles. Deléage was one of the regional historians in France and he published a major work on Burgundy in 1941–1942. He worked with Bloch and paved the way for a new understanding of the feudal system. See Bloch (1953) p. 96, 206 and Bloch & Dauvergne (1956), p. 128–132, and also p. xv, xlii, 9, 18, 50.

are good. Maps of the field layout can sometimes be reconstructed down to the fourteenth century.

He was quite proud of his results, and, later in a text about himself, he declared that one of his major contributions was that he had created a method for measuring medieval villages and exploring European land systems.[71] I have not found any quotations to these articles in European research, and very few in Swedish. Perhaps, had he had the opportunity to continue with this approach, he would have garnered more long-lasting success. His two Swedish books from 1947 and 1951 were reviewed in international journals (a subject to which I will return).

Folke Dovring was trained in the military service in geodesic measuring and he had previously produced some maps for his books about Sweden. But, in these articles about European villages, he excelled in producing large and detailed maps. Later he mentioned that one of these maps took four months to reconstruct and draw. He tried to establish the actual area of the oldest measurements to be able to date them, or at least see possible relationships between them.

He is careful with conclusions, and talks about the need for more detailed village studies before any well-founded conclusions can be drawn. But he hints, especially in the article about Normandy, that the land measuring system could have been developed early on as a mean to distribute the duty to serve the royal army in the Carolingian empire. He then connected this with his theory about the earliest Swedish land measure systems, and proposed that these means to distribute duties could have set an example for and been spread to England and Scandinavia.

His results are much in line with later research in European medieval and agricultural history, but Dovring's results are seldom mentioned by the leading scholars.[72] Perhaps this would have been different if he had continued with this field of research.

For the time being he experienced success. During the summer of 1950, he went to a congress in London where he presented his work on Grotius to a select audience of lawyers and historians of law from around the world. And thereafter, he went to the world congress of historians in Paris where he presented his *kasuistik*-method. "More empirical studies, less speculation was the theme I had chosen", he told his mother in a letter.[73] He considered his speech a success. Indeed a

71 Dovring (1968).

72 The "mansus" is treated by all the agricultural historians presenting on the Middle Ages in the post-war era, for instance Postan (1972) in England, Duby (1968) in France, and Rösener (1992) in Germany. However in his supplementary volume on Marc Bloch's rural history of France, Robert Dauvergne refers to Dovring's article in *Annales* 1951, Bloch & Dauvergne (1956), p. xl.

73 PA-MS, Folke Dovring to Naemi Ossiannilsson, 17.12.1950.

good measure of success is that he was invited to publish it in *Annales*, the most famous of all historical journals. He gave the presentation of the method there in 1951, and was the second Swede to publish in this journal following internationally renowned historian and economist Eli Heckscher.[74]

Folke Dovring also, in the letter to his mother, told the following interesting story. "After my speech I was resolutely taken hold of by a slender and red-haired man, who I'd met before, and who wanted me to have a chat and a cup of coffee with him. It was the Cambridge-professor Postan, one of the leading scholars in economic and social history." Dovring comments on Postan's life; born in Turkey, he studied in Russia and escaped to England after 1917. Dovring continues: "He holds that my casuistry method is the only correct method for history of agriculture and he sincerely asked me to come to Cambridge as soon as possible."

He was 34 years of age, and destiny seemed to smile on Folke Dovring.

74 Odén (1978), p. 19.

CHAPTER 3
On the academic battlefield (1951–1953)

Selection and exclusion

Dovring must be considered as one of the most successful young historians in Sweden around 1950, and yet he was pushed aside. He applied for a chair in history and was not only put down in the concurrence, but also officially declared by the majority of the evaluators as not competent enough to be a professor of history. Such declarations were unusual. It meant that all doors to a position at the universities in Sweden were closed to him. If he wanted to stay in the field, his only opportunity was a career as a high school teacher.

If the question can be lifted to a more general level, one can ask: *What does it take for a man or a woman who is very intelligent and hard-working to make himself an outcast from the academic society?*

Leaving lack of talent out of account, the reason why someone is excluded from the academic society can be sorted into four subheadings: ideology, personality, relations, interpretations. I will mainly concentrate on the two last-mentioned.

Persons with unaccepted political standpoints have less latitude in most societies, but usually scholars are given more freedom than others. The idea of academic freedom has had a fairly strong position in Europe. After World War II, former fascists were expelled on a grand scale from many European universities, but many of them were later allowed to return. In Sweden, the purge after the war was less dramatic; even Nazi sympathizers, such as the professor in history Gottfrid Carlsson in Lund, were able to keep their chairs. There was, however, a considerable restriction of influence for such persons.

Folke Dovring was not himself known as politically active. As a son of the former Mussolini adherent K.G. Ossiannilsson, Folke could have experienced disadvantages. But too many were connected with people who had been on the wrong side in the war for this to be a plausible explanation. Besides, Ossiannilsson had changed his political position during the war, and supported the allies.

Personality plays some role in explaining a career. When I talked with people who knew Folke Dovring they have given different opinions. Torsten Husén from the group of friends he had in Lund, considered him in possession of charm and social capacity.[75] His fellow graduate student Gunnar Westin remembers him as

75 Interview with Torsten Husén, 25.3.2002.

isolated from the inner circle in the department of history.[76] Karin, his wife who knew him so well, explained in a letter to his mother in 1949 that "Folke is so self-critical, and hardly never wants to meet people if he does not feel fit and if he thinks he cannot give them something substantial in their discussion".[77] Later, at the University of Illinois, he was not inclined to small talk, but willingly helped students who came to see him.

The only physical disability Dovring had was a small tendency to stutter, which he mainly overcame as a student in Lund, where he took lessons with a speech therapist. This speech impediment is very seldom mentioned in the interviews, and, even though it may have caused him to hesitate in oral discussions, it would not have been of any hindrance to his scholarship.

Intelligent, but eccentric, professors populate the hallways and libraries of any university of standing. Only a dysfunctional academic milieu would try to get rid of such scholars and let other, more conforming persons, rise to the top. A slight reticence in Folke's personality would probably have explanatory value only if he were in danger of being excluded because of other reasons.

Every conflict has an agenda of contacts and relations. Scholars tend to form a hierarchy and most would know their own position. Discussions in seminars, at conferences and in reviews will settle and resettle this hierarchy – but other factors, especially personal relations and subordination, will also play a role.

Dovring felt that he was not valued highly enough in the department of history in Lund, and decided to glean support from some group outside the established hierarchy. He persevered and began to receive international acknowledgment for his research. Such a strategy is dangerous however; counter-reactions based on jealousy might come to the surface.

In an academic society it is generally accepted that some will rise above the others, but envy, as a group feeling, could be unleashed if an established order of priority to higher positions is put into question – especially if the established leader feels that he is threatened. This is what happened during the battle in 1951–1953. Other scholars felt that Dovring tried to push himself forward in the queue, and he did this by insulting the organizers of the queue.

The role of new interpretations should not, however, be underestimated when a scholar has difficulties. To have an interpretation that is different from the mightier among the tenured professors could be risky, but one must take into account that young scholars often are supposed to be oppositional. Many professors would favor a young student who scrutinizes his supervisor's old thesis, because then the professor will feel that he and his ideas are still of interest. Certainly some professors prefer obedient students, but thereby they often expose their research tradition to the eventual death of boredom and mediocrity.

However even an open-minded professor wants the opposition to be moder-

76 Interview with Gunnar Westin, 3.2.2002.

77 PA-MS, Karin Dovring to Naemi Ossiannilsson, 27.9.1949.

ate; a total onslaught on the supervisor is seldom appreciated. Preferably, the opposition should point to a further development and amendment of the body of ideas in which the graduate student has been trained.

Preparation

Dovring was an uncompromising candidate as he had been an uncompromising son. Among the graduate students, professor Sture Bolin's favorite was Sven A. Nilsson, who was well-established in the department when Folke Dovring came. In his group was also Sten Carlsson, the son of professor Gottfrid Carlsson, the other professor at the department.

During his years as an undergraduate, Folke Dovring received fewer grants than many other students. He had found it necessary work halftime as a librarian for nearly all the years he labored on his doctoral thesis. That he and Karin had to live under poor conditions partly explains why he engaged in conflict with his father over financial support.

It was always difficult to get grants, but a reason why Dovring had more difficulties than others could have been what Dovring later himself wrote: "Already in an early phase of my research pronounced differences of opinion appeared between professor Bolin and me considering central issues."[78] According to Dovring, the differences were caused by Bolin's ignorance of the sources, which he probably told his professor. Dovring reported that their discussion often ended with Bolin saying that it was possible that he, Dovring, was right, but that he himself, Bolin, did not believe it.[79]

The defense of a thesis, the "disputation", in Sweden is somewhat old-fashioned. The thesis is published three weeks before the defense, so all can read it. The defense then occurs in public before an audience that can consist of relatives, other students, professors from different fields, etc. These occasions are entertaining shows in the life of a Swedish university, and attending them is considered as a part of the education of the graduate students (they should know what awaits them). Around 1950, these occasions could go on for half a day or more with at least two opponents selected by the faculty and extra opposition offered from the audience. Folke Dovring's main official opponent was Gunnar Westin, one of his fellow graduate students at the department.[80]

Westin never published his opposition, which is a little bit strange as the custom is to publish every opposition as a review in a leading journal. This is considered as a part of an ongoing quality control. Therefore, it is impossible to know what he said during his opposition. When I asked Gunnar Westin why he did not publish his opposition he answered, "We did not understand his methods. Years

78 Dovring (1952), p. 5–6.

79 Dovring (1952), p. 6.

80 PA-MS, Karin Dovring to Naemi Ossiannilsson, 27.9.1949.

later, when I started to work with the same sources as Dovring had worked with, I learned to appreciate him and his methods much more."[81]

By strange circumstances, the copy of Dovring's dissertation that Westin used when he prepared his opposition has fallen in to my hands. Dovring wrote on the first page: "to my dear friend docent Gunnar Westin" (in Swedish: "Docent Gunnar Westin från tillg. vännen Respondenten"). This friendship was not to survive the disputation. My Ph.D. student Alf Ericsson, who is doing his research about the attung and other medieval land measure units, found this copy of the book and handed it over to the archeological investigation unit where he was working. Later, when we, Alf Ericson and I, started to work together, he showed it to me, but this happened so late in my research on Dovring that I could not present the copy to and discuss it with Gunnar Westin, who passed away 2008.

I have gone through hundreds of red marks in the margin and tried to make some sense of them. Mostly these notes consist of questioning details and recounting the results. Westin is suspicious of the *kasuistik*-method, and this seems to have been the main critique throughout all the notes. He considered Dovring's book hard to read and understand, a note says "unreadable", which is "oläslig" in Swedish. A few times Westin remarked in the margin that Dovring misinterpreted Lönnroth, but this is not a major theme in his notations.[82] In the interview Westin thus seems to remember rather correct what happened. The method stood in the center of the discussion, and on that point Westin later changed opinion.

Dovring's two major books were never reviewed in the two leading historical journals in Sweden, *Scandia* and *Historisk tidskrift*. But his thesis from 1947 was reviewed in the leading historical journals in England, Germany, France and even in Finland. This makes Dovring's thesis nearly unique in Swedish historiography, with a silence in the native country and more attention abroad than most other young historians.

In *The Economic History Review* 1949 Michael Postan gave an account of the main results of the thesis. He thought that it was an important contribution to show that the Swedish system probably was an outflow of a general European change. That Postan approached Dovring at the conference in Paris was thus not by chance, he wanted to discuss with a young scholar that had impressed him.

In Finland the leading historian Eino Jutikkala presented the thesis in the national journal of history, *Historiallinen Aikakauskirja* 1950, and was generally sympathetic. He also pointed to the fact that Dovring had shown the European connection. The French historian P. Jeannin described the thesis in *Revue historique* 1951, and he concluded his enthusiastic review with the statement that it is a pity that the language prevents: "une discussion du livre par des specialists français de l' histoire agraire". The German review by Hermann Kellenbenz came in

81 Interview with Gunnar Westin, 6.6.2002.

82 Gunnar Westin's copy of *Attungen och marklandet*, National Heritage Board, Department for Archaeological Excavations: East (Linköping, Östergötland), Office library.

Vierteljahrschrift für Sozial- und Wirtschafsgeschichte 1954. It was delayed by the reconstruction of the academic life after the war. Dovring was aware of this review in 1950 and he mentions it in a letter. When it was published, it was very complimentary.

At the dissertation in Lund in springtime 1947, praise was not that extensive. Dovring's widow Karin told me that Folke thought Westin was unjust and harsh, and her memory is confirmed by a contemporary letter from her to Folke's mother (see below). After the public part of the dissertation, the evaluating committee came together and assigned a grade. The minutes from the committee have been preserved. Dovring's supervisor, Sture Bolin, posited that the thesis had some qualities, especially in the way Dovring treated and presented the sources. But Bolin also remarked that Dovring tended to over-interpret the sources and Bolin especially argued against Dovring's attack on Erik Lönnroth.[83] Nonetheless, Dovring got a high degree on his dissertation. As a result he could readily, without presenting further publications, apply for the formal degree of "docent". The importance of this shall immediately be revealed.

In Sweden around 1950 the next step in an academic career after the dissertation would normally be to obtain a grant for six years, called a "docent-grant". It was not possible to apply for such a grant without the formal degree of "docent", and obtain a "docent-grade". Being a docent it was possible to get a docent-grant. This six-year grant was comparable to the position of assistant professor in the U.S., but after the six years, no automatic evaluation would have taken place. Instead, a docent had to wait for one of the full professors to retire.

The number of chairs was restricted. Around 1950 Sweden had seven professorships in history at four universities: two in Uppsala; two in Lund; two in Stockholm and one in Gothenburg. The last professor to be appointed before the early 1950s was in 1942, and thereafter followed eight years when no new professors were appointed. A growing number of docents was waiting and preparing for a period of competition. From 1950, for half a dozen years, several chairs became available for competition. (Later, in the 1960s, the number of students and professors increased; and today the whole Swedish system has become more similar to the American.)

It was recommended, probably by his supervisor Bolin, that Dovring should apply for a docent-grade somewhere else than at the department of history at the College of Liberal Arts in Lund. Dovring chose to try for a docent-grade in the subject "agrarian history", which was a new subject, at the law school in Lund. This caused some confusion and even protests among the professors in law, but he was eventually accepted.[84] For a time, he planned a career as a historian of law and

83 DA, Excerpts from minutes of Humanistiska sektionen, 28.2.1947/ utdrag av Humanistiska sektionens protokoll den 28 februari 1947.

84 DA, Excerpts of minutes from Juridiska fakulteten in Lund, 21.3.1947/ utdrag av Juridiska sektionens protokoll den 21 mars 1947.

started to take courses at the law school.

However, half a year after he presented his thesis, he applied for and was awarded a docent-grant in his old department of history. He obtained it at the same time as Sven A. Nilsson, the favorite of Bolin. In the summer of 1947, both these young doctors started a race. Who would win – the one who produced the most or the one who was more liked by the professor? Dovring's strategy was to produce as much as possible, and to publish abroad so he could escape the judgment of his own department.

The combatants

From the 1920s to the 1950s, an internal strife brewed in the society of Swedish historians between a quite well-organized group, the Weibulls, and a more loosely-organized group of "conservative" historians. The first-mentioned group was labeled after its core, the Weibull-family: the father Martin Weibull and his first-born son Lauritz Weibull, who became the leader of the group, and the younger son Curt Weibull, who was a fierce fighter. All three held chairs in history in Lund and in Gothenburg. Besides advocating new ideas, they also launched a new way of educating Ph.D. students. Instead of leaving the postgraduate students more or less alone to finish their thesis, the Weibulls started regular seminars. Texts were discussed and penetrated, and thereby a group feeling took form. In such a group, the internal hierarchy can be strong, especially if the leader advances his favorites. This certainly happened in the Weibull-group.

The other group partly consisted of followers of the charismatic conservative historian Harald Hjärne, who dominated the scene around 1900, but after him no new natural leader had emerged. In addition two of the leading professors, Sven Tunberg and Nils Ahnlund, disliked each other.

In a recent doctoral thesis Håkan Gunneriusson has studied the two groups and their struggle over positions from a sociological point of view. As source he has mainly used preserved letters. With information from Gunneriusson's thesis, I have tried to describe the historians 1950–1954 in a sociogram. This must be seen more as a picture than as a precise description about every possible link between participants. In particular, the thickness of lines and arrows are more tentative than exact. Much of what happened did so in secret, and oral contacts were decisive in many of the outcomes. Below I use the acronyms I also use in the sociogram, as (fd) for Dovring.

As a complementary source I have used the forewords, the acknowledgments, in the printed thesis of the combatants.[85] They where formed as a declaration about acquaintances, and give a hint about connections established during the years as Ph.D. students, before the real competition started. A formula had to be

85 In Myrdal (2008) I have analyzed all the forewords written by historians taking part in the competition for chairs during these years.

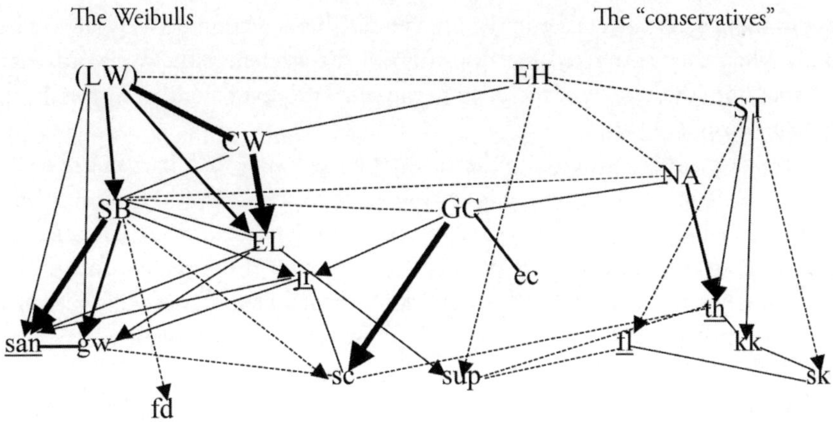

Figure 2. Sociogram of the main participants 1950–1954 in the concurrence over chairs in history.
Legend:
SB = The professors up to 1950 have initials in capital letters.
(LW) = The retired but still active player is placed in brackets.
fd = The "docents" have their initials in lower-case letters. I have included those who applied for at least two different chairs and those who tried in an earlier period, i.e. 1942 and again 1950.
san = The winners, who got chairs, have their initials underlined.
⟶ = Arrows show support and domination.
—— = Lines show support and friendship.
The thickness of the arrows and lines varies. Dashed arrows/lines stand for more formal contact, as for instance supervising a doctoral thesis. The thickest arrows/lines stand for very close contact, as being related to someone as a son.
Legend, names: Lauritz Weibull (LW), Erland Hjärne (EH), Sven Tunberg (ST), Curt Weibull (CW), Sture Bolin (SB), Gottfrid Carlsson (GC), Nils Ahnlund (NA), Erik Lönnroth (EL), Jerker Rosén (jr), Einar Carlsson (ec), Sven A. Nilsson (san), Gunnar Westin (gw), Sten Carlsson (sc), Sven Ulric Palme (sup), Folke Lindberg (fl), Torvald Höjer (th), Kjell Kumlien (kk), Salomon Kraft (sk), Folke Dovring (fd).

Sources: Gunneriusson (2002), generally also Odén (1991) and Svensson (1994) on Hjärne (EH) and his relations.

followed, where the professors must be mentioned first, but, following this, the graduate student had a freedom of choice in mentioning comrades and colleagues. The closeness of the Weibull-group is obvious in the forewords, but additional information may also be gathered. The acknowledgments show that Sven A. Nilsson (san) and Dovring (fd) had contact during the years they worked with their thesis as they thank each other in the forewords. This acquaintance was later to be broken when they competed for a position in the academy. The forewords also show that both of them were included in the small group of agrarian historians in Lund at the time.

The sociogram is restricted to the years when Dovring took part in the competition, 1950–1954. During the following years, Gunnar Westin (gw) and Sten Carlsson (sc) became professors and several years later also Sven Ulric Palme (sup). By then new and younger players had also entered the stage. Most of the participants in the sociogram who were not awarded chairs eventually received the title of professor, which is not at all as prestigious or prominent as having a chair.

The Weibulls were a closer-knit unit than the "conservatives". Erland Hjärne (EH), the son of the famous Harald Hjärne, had an intermediate position, but was not heavily engaged. His student Sven Ulric Palme (sup) tried to jump from one group to the other by establishing contact with Erik Lönnroth (EL). Other combatants such as Kjell Kumlien (kk) and Salomon Kraft (sk) were placed low in the evaluations as they were under attack from the Weibulls and could not get strong support from their allies.

In this sociogram Folke Dovring (fd) is left rather alone, with weak support from his supervisor, but at the same time, under attack from Erik Lönnroth and his close benefactor Curt Weibull (CW). From a strict point of academic sociology, Dovring did not belong to either major group. Around 1950, when the long expected showdown took place and new professors were appointed to several chairs, Dovring was pushed aside in an early phase of the battle to clear the field for the two groups.

It would be too simplified to see this network as the most important factor in sorting scholars. A gifted historian could succeed even though he did not have very strong support, as did Jerker Rosén (jr). And Folke Lindberg (fl) got a newly established chair although he had weak support from historians at the national level. Instead Lindberg had strong support from economic historians outside the sociogram and from the political level, namely the Stockholm city-counsel.

Different goals

The real conflict was, however, not about intrigues and positioning from individuals and groups, it was a conflict over historical methods and interpretations.

The Weibulls, who called themselves the source-critical school of historians, wanted to sharpen the scrutinizing of sources, and they wanted to emphasize other motives for historical change than that of the building of a Swedish nation.

The other group did not name themselves "conservatives", this was a stamp put on them much later by Swedish historiography, which has been dominated by the followers of the Weibull group. This other, more loosely knit group, were only partly connected by their political ideology concerning the state. Even some of the Weibulls, such as Sture Bolin, were politically conservative. However all of the Weibulls were politically against the Nazi-regime in Germany. By contrast, some of the "conservatives" – but far from all – admired the new regime south of the Baltic. Actually, the accusation of "Nazism" was readily hurled as a means to undermine persons who belonged to this group but who were actually outspoken against Nazism. For instance, Erik Lönnroth did this in as early as 1935 against Nils Ahnlund, a leading professor in the other camp.[86]

The source-critical method, given the meaning as used in Europe as a whole, was introduced among Swedish historians in the late nineteenth century. Sources, especially from the Middle Ages, were published en masse, competently and keenly. Historians at the time accepted that sources that are the most reliable are those that are the least tendentious and are close in time and space.

Peter Burke has described how the new source-critical method tended to strengthen concentration on the history of the state (see foregoing chapter). As one could expect, this also was the case in Sweden. Historians in the first group of source-critical historians, centered around Harald Hjärne who dominated the field around 1900, were critical of cultural history. Hjärne and his followers contended that the history of the state must be the center of historical research.[87]

The Weibull school wanted to go one step further than the first source-critical group of historians, and totally clear away all unreliable sources. Neither cautious use of suspicious sources nor reconstruction by combining different problematic sources were allowed. The goal was to establish an absolute historical truth built on the few remaining reliable sources, after the unreliable had been purged from the discussion.[88]

All historians did not share this belief in absolute historical truth. The Weibulls tried to distinguish between "less reliable narratives" and "more reliable vestiges". Harald Hjärne, who had introduced the concepts of "narratives" and "vestiges" among Swedish historians, explained that a source could be a "narrative" in one context and a "vestige" in another and furthermore he claimed that truth could never be exact.[89] These arguments would later echo in Folke Dovring's *History as a Social Science* from 1960, which I will treat in the next chapter.

Expelling sources was used as an efficient method against nationalistic myths, but there is a temptation in this method. That is to sort out sources not in favor of the hypothesis the scholar wants to defend. And so it came to be used. The "conservative" historians often turned out to be more correct in their attempt to master the totality of problematic sources than did the over-critical Weibulls who

86 Gunneriusson (2002), p. 80.

87 Torstendahl (1964), p. 277, 294.

88 Thorstendahl (1964), p. 246, 351–357, 365; Odén (1975), p. 149, 202, 207.

89 Torstendahl (1964), p. 372–373.

tried to rely on a select number of sources. The focus of history did not change. The strict source-critical method was best suited for discussing single events. Had a person barely visible in the early sources really existed? Was the famous battle mentioned in every schoolbook perhaps a fake?

In their first attack in the 1910s, the Weibulls, especially Lauritz Weibull, cleared Swedish history from many nationalistic myths. From an ideological standpoint, this was combined with liberal ideals. Idealization of the Swedish nation was not a goal as such. The ideology of the Weibulls was never clearly stated or built into a theory. This is understandable as it would have been contrary to their self-concept as a group who had found the method to establish the real truth.

Besides discussing state building, the Weibulls also emphasized the role of trade.[90] This was both in accordance with their liberal ideals, and followed an international trend in the early twentieth century. But still it was the role of trade in politics, the state and the nobility that was the focus of their interest.

They had contacts with Danish scholars and were united with them in a common denial of nationalistic myths about the historical struggle between these neighboring countries. But, other than to a surprisingly small extent, the Weibulls did not reach out to the larger international discussion, and their method was not widely used outside Scandinavia. Bolin was of course the exception, with his broad international network and his research on European issues – but he was also less of a real Weibullian in his methods than the others.

When the definite breakthrough for the Weibull group came in the early 1950s, the aims of the whole international historical society had changed (see foregoing chapter). Instead of singularities, the large historical structures and ordinary people came into focus. The early Weibullians with their close concentration on the single correct source was not very well adapted to this new research trend. Quantitative history with its combination of sources and building of reconstructions took over. Dovring thus was right but in the wrong time and at the wrong place.

The leader

A new dimension was introduced in the Weibull school by Erik Lönnroth in the 1930s. He pointed to the economic basis for state building.[91] Later he developed this into a theory, not clearly outspoken however. On every historical question, he posed the questions about the self-interest, especially the economic self-interest. Every action could be explained by the simple question: who gained in economic terms? This was something totally different from the nationalistic ideas, which

90 Odén (1991), p. 160.

91 Torstendahl & Nybom (1988), p. 87–88.

formerly had dominated the interpretation of Swedish history.

Lönnroth combined this new idea with a talent for presenting fascinating, plausible but also provocative hypotheses. He could present his thesis in a most eloquent style. Such scholars are highly needed in an academic society, as they trigger others to testify the truth and open up new fields of research. A well-known example is the French historian Philippe Ariès, who eloquently presented a hypothesis about the non-existent childhood during the Middle Ages. Children as a category, he claimed, was something that came late in history. His hypothesis has been proven many times to be invalid, but the ways in which this has been shown has given us a totally new insight in how children lived and behaved in the Middle Ages.

This imaginative talent of Lönnroth was combined with the strict source-critical restrictions advanced by his teachers, especially Curt Weibull; this proved to be a very fertile mixture. Lönnroth's first two books, and a couple of his articles, presented a new view about how the Swedish state apparatus and nation emerged and changed during the Middle Ages. He based his hypothesis on diverse sources, which he presented and utilized. For many of his contemporaries, his way of presenting and his interpretation, gave a new meaning to the study of history.

A general feature in his interpretation was that he emphasized the role of trade and commerce, from the Viking age through the Middle Ages, another that he underestimated the dynamic role of agriculture. To him every strife or leading group seemed to have been centered on commercial interests, be it Viking Age magnates or high nobility in the Late Middle Ages. The role of the common man was not recognized.

The Weibullian emphasis on trade was in accordance with a European trend in the 1920s and 1930s. Trade as a dynamic factor had been successfully advocated by the Belgian Henri Pirenne. (He was more interested in general economic development, than in the specific question of state building.) In the 1950s, an opposition against the overestimation of trade came forward, and an interpretation emphasizing rural history came to dominate among the medievalists in Europe for a long time (Dovring was a part of this movement). But Lönnroth would never sway in his belief in the decisive role of trade and still proposed such interpretations many decades later.

Lönnroth's skill at formulating theories was hampered by his tendency to let his attractive hypothesis carry him away and to direct his interpretations of details in favor of them. He was not a meticulous researcher who collected a mass of reliable facts on which others could build their theories. Several misinterpretations have been proven. For instance, a fierce discussion went on in the 1930s about the large rebellion in Sweden in 1430s, and Lönnroth, in a typical Weibullian way, tried to discredit several of the main sources. His dating and attribution of these sources has turned out to be wrong.[92]

92 See Schück (1980).

It is interesting that this question, of Lönnroth's accuracy in his detailed interpretation of the sources, seldom is posed in modern historiography.[93] This forgiveness of the brilliant Lönnroth was a fact early in his career, and Folke Dovring never ceased wondering about it. Time and again Dovring reflected on how someone that is considered a genius was allowed to present theories and ideas without factual base.

Lönnroth became one of the mightiest historians in twentieth century Sweden. This said, he does not deserve the kind of hagiographies that have been written up to now.[94] He should be given a real scientific and critical biography pointing out his strength as an imaginative and provocative historian, but also discussing the most interesting question about how he managed to establish a very strong position of power in the society of Swedish historians.

Lönnroth held a chair in Uppsala from 1942, and then the chair in Gothenburg from 1953. He held many important positions: he was the chairman of major funding organizations, he worked for the government in organizing higher education, he sat in the governmental board for higher education. He crowned his career when he became one of the eighteen members of the Swedish Academy, presiding over the Nobel Prize in literature.

He used his high position to direct, exert influence and manipulate opinion. I will just give one example. Erik Lönnroth was for many years the chairman of the most important of the governmental funding organizations, the Swedish Council for Research in the Humanities (later enlarged by including research in social sciences, but still under Lönnroth's chairmanship). In this organization he was also the chairman for the committee evaluating the applications in the field of history. The story, which I have checked with several persons, goes like this:

Erik Lönnroth comes to the meeting, and the three other professors of history in the committee are already waiting. Lönnroth then fishes a sheet of paper from his pocket, and reads aloud who is going to get grants. He has not talked with any one in the committee beforehand, and he does not consider discussion at the meeting necessary. Besides, he is in a hurry to get to his next meeting. The meeting is concluded in half an hour, and the committee decides what Lönnroth has suggested. There is some grumbling from the other professors at the lunch they have together afterwards. But they accept Lönnroth's decision not only because they are subordinate, but also because they consider it as a fairly good evaluation.

He also thought of himself as one who could glean money for large projects in history from unwilling politicians. In a retrospective article written late in his life, he presented the story as he saw it.[95] The right-wing nationalist historians had

93 Gunneriusson (2002); Englund (2002).

94 As for instance Englund (2002), which however was the obituary for Lönnroth as a member of the Swedish Academy, and thus close to the genre of hagiography.

95 Lönnroth (1998).

dominated for a long time, so when social-democrats and liberals came to dominate the political stage, they wanted to punish the historians. Lönnroth complained that he and his colleagues were being "punished for our predecessor's right-wing ideology".[96] However the governmental funding organizations were spared from this reduction. Therefore, only "through the activities of the Swedish Council for Research in the Humanities" and efforts by some professors, was a certain amount of funding for history effected.

He used his position of power to reconcile the two fighting groups of historians, and he helped many skilled young historians. He was able to raise considerable grants for historians, and, not the least, for medievalists. Never the less, he was also harsh on graduate students or other scholars who dared to criticize his theories. The distribution of grants and funding was by no means spread to people who had opinions he strongly disliked.

Folke Dovring, who had attacked him so much, was probably someone that Erik Lönnroth did not appreciate. It must be said that Lönnroth was seriously interested in a general armistice in the society of historians. When he acquired the position of the leader, the "prince", of historians he wanted the society of historians to stand strong against politicians and other stakeholders.

Karin Dovring maintains that Folke Dovring on one occasion met with Erik Lönnroth, who had suggested that they should accept the fact that the truth could be somewhere in between and work together. Karin was not present at the meeting, and her account of what was said could be unreliable, but according to this account Dovring at the time apparently felt that Lönnroth had stretched out a hand.

Close to the battle

During the spring of 1951, something must have happened. We know that Dovring then decided to change direction in his research from the Middle Ages to the modern era. It was half a year after his splendid success abroad in the summer of 1950. Perhaps his professor, Bolin, had told him that he would not support an international career in medieval history. We never will know, because such messages are not written down, and perhaps not even verbalized clearly. Bolin certainly had a very strong position on the international arena, and could hinder (or advance) any of his pupils' careers. People abroad would listen to Bolin's opinion.

In a long letter from Karin to Naemi written in April, 1951, the question of envy is addressed.[97] She worries about her own approaching defense of her thesis. Then she refers to how badly treated Folke had been when he presented his thesis, though he later became renowned among European scholars. "Here at home they

96 Lönnroth (1998), p. 50.

97 PA-MS, Karin Dovring to Naemi Ossiannilsson, 17.4.1951.

still do not understand his methods, only that he has become famous! It is not fun at all." And she tells her mother-in-law that the university treats her and Folke as if it had "a lot of brilliant people" to choose among.

In May, 1951, Karin Dovring presented her thesis in Lund; it was barely accepted.[98] She was downgraded mainly for the same reasons as Folke – her methods were not understood. Later Karin Dovring's work gained an international reputation in research about propaganda.

It was at this time Dovring started to look for a position elsewhere, and he tried to exploit the achievement he felt he had made abroad. He got a grant to study recent history of agriculture and started to travel in Europe to collect material.

He also in July, 1951, wrote to the leading professor in history in Uppsala, and asked for a position there.[99] A "chair professor" at that time in Europe had quite an important position as leader of the institute. This professor happened to be Erik Lönnroth who Dovring earlier had criticized. One can wonder why Dovring wrote this letter, but, in the letter, he talked widely about his success abroad to make a favorable impression.

Lönnroth answered in a friendly tone, and said that no position at all was available.[100] Then Lönnroth made the suggestion that Dovring should try "to make some connection with Ultuna (the College of Agriculture outside Uppsala) and persuade them to establish a position as teacher in agrarian history?" Lönnroth started to talk about strategies; he was a keen player of such games. He mentioned the possibility of approaching the politicians, especially the "Farmers party" ("Bondeförbundet") in the parliament. Lönnroth also mentioned proposals for a chair in rural sociology ("lantbrukssociologi") put forward at this time at Ultuna (it was never realized). The College of Agriculture had at that time an anti-humanistic profile, and the faculty would not have accepted a chair in agricultural history. Lönnroth was very clear about that, and states that the natural scientist will not be interested, but the political way could be a possible way. He also mentioned that they would have a possibility for further discussion at a conference they both soon would attend, probably the one that Karin remembered Folke talked about.

One complication, of which Lönnroth perhaps was not aware, was that Folke Dovring's eldest brother, Frej Ossiannilsson had held a position at Ultuna for two years. Certainly Dovring did not want to work at the same place as his brother, whom he disliked so much. He also did not want to play the academic-political

98 See Landgren (2005), p. 128–131, 158. Karin still believes that she was discredited by her acquaintance with Folke, but the discussion in the examination committee was on her methods and results, especially her use of methods from social sciences (propaganda research) in literature.

99 DA, Folke Dovring to Erik Lönnroth, 8.7.1951.

100 DA, Erik Lönnroth to Folke Dovring, 11.7.1951.

game that a new chair in agrarian history would have demanded. Probably he suspected that he was actually to be made a sacrificial pawn. Later, when the conflict had evolved into pamphlet writing (see below), he declared that agrarian history ought to belong to the subject of history. This was probably an indirect answer to Lönnroth's earlier suggestion.

But Lönnroth's offer of support for a chair in agrarian history could have been more than a formal gesture. Perhaps he seriously wanted to help Dovring, even though Dovring had attacked him. During the years 1948–1950 there had been an attempt to form a society for the history of agriculture, or a research institute, with the support of the farmers union.[101] This had failed, but from Lönnroth's perspective it could have been worth a new try, had Dovring been more receptive. Lönnroth had the political connections, and, had an attempt succeeded, the Weibulls would legitimately have gotten rid of a dangerous competitor. A new chair would also have strengthened the position of history research in general. Moreover Dovring would have owed a debt of gratitude to Lönnroth. This is, however, sheer speculation; what we know is that Dovring did not at all consider this suggestion.

Immediately after he got Lönnroth's answer, Dovring wrote another letter to the leading professor of history in Stockholm, Torvald Höjer, and got the same negative answer; there was no position available for which Dovring could apply.[102]

What Dovring considered as the real opportunity for which he had waited, instead came in the autumn. Dovring wrote to his mother that all of August he had worked on his application to a chair in Uppsala, as the professor Erland Hjärne, who occupied this chair, suddenly decided to step down earlier than Dovring had expected.[103]

This was the second of a series of competitions for chairs during these years. Dovring had not applied in 1950 when the chair in Stockholm was advertised. Now, in Uppsala 1951, he considered himself competent enough to try. The ranking would decide his future chances. Perhaps he also thought he had a fair chance of obtaining this specific professorship, considering his international success the foregoing years.

Academic conflicts

Appointing a new professor in Sweden was a complicated affair. When a professor for some reason, usually retirement, left his chair, it was announced, and competent persons submitted applications. The first step in the process was to appoint

101 Odén (1991), p. 145, 165.

102 DA, Folke Dovring to Torvald Höjer, 15.7. 1951; DA, Torvald Höjer to Folke Dovring, 17.7.1951.

103 PA-MS, Folke Dovring to Naemi Ossiannilsson, 7.10.1951.

an evaluation committee with three or four members. Suggestions from other universities were collected, and then the university where the chair was situated chose the committee.

Around half a year later, the committee presented a ranking, which was made public together with extensive comments by the members of the committee. These evaluations were important instruments in the internal ranking within the society of Swedish historians.

Then the faculty of the college discussed the matter, often with minutes made public for everyone to read. Next the faculty of the whole university voted. The following step, in the 1950s, was the decision of the chancellor of the university. The last step was the government who had to confirm every professorship in the country.

Every one of the candidates had the right to file complaints. Often these protests were published and would even develop into small booklets spread by discontented candidates. This genre, the "apply pamphlet", was an important part of the academic dispute in former days. Today it has totally vanished. A well-known Swedish author, Frans G. Bengtsson, once declared that academic apply pamphlets were among the most amusing literature one can read, often with outstanding qualities of epic and passion.[104] Bengtsson was an outsider to the academic community. To those engaged in a conflict, it was not that easy to joke about them.

Often non-academics tend to wonder about the heat and ferocity with which conflicts at the university are fought, and many feel that there is something unreasonable tied to such conflicts – i.e. that dispute is a goal in itself, that professors are always quarreling, that it is their lifestyle to do so.

In any workplace, conflicts can be destructive for those involved, but the academic controversy is often more public. From one angle, the intellectual conflict is an integrated part of the academic life; without academic disputes, scientific research would perish. Ideas are born and die in a selection through open discussion. But academic conflicts entail much more than that. In the battle, the fundamentals of the combatants' intellectual capacity will be threatened; for many academicians this is the foundation of their life.

Academicians can be compared with actors or authors. An author will put his or her personality in the book, and, when it is sent out, the reviewers can crush his book or applaud it. Normally authors are more engaged by the bad reviews, and tend to look at reviewers as people who do not really understand their books. This animosity, and, at the same time, symbiosis between critics and writers is essential to the function of a literary society. In the academic society there is no distinction between writers and critics; all members of this society are both, which gives many more possibilities for revenge than with authors or actors. This heightens the conflicts.

---

104 Bengtsson (1938), p. 264.

The ferocity in the competition for the highest positions in the academic arena can also be explained by the advantages afforded in such positions. The life as a professor is quite pleasant. You are interested in something that you would probably have as a hobby if you did not have it as a job. Other advantages are a lot of disposable time, a high status in society, and a salary above the average.

More important still, is that a professor not only perform research in a subject but also shape and form the field of research; that is what most successful academics dream of accomplishing. Academic conflicts may have many causes, but ideas and interpretations will always be in the core of the conflict.

The battle phase 1 – turned down

The appointed committee for the chair in Uppsala in 1951 consisted of three professors: Dovring's former supervisor Sture Bolin, Erik Lönnroth's former supervisor Curt Weibull and a Norwegian professor of history, Sverre Steen.[105] It took nearly a year before they presented their report. It was made in public and Curt Weibull even published his evaluation as an article in *Scandia*, one of the two leading journals of history in Sweden, and the mouthpiece of the Weibull group. To publish evaluations in journals was unusual, but Curt Weibull had done so before. For him this was a part of the struggle between the two main groups of historians.

Among the five candidates, Folke Dovring was not only ranked lowest, he was actually considered by two of the members of the committee as unqualified to be a professor in history. In Swedish the word "inkompetent" is used, but the meaning of is not as harsh as the English "incompetent", it would rather be translated as in between the meaning of incompetent and unsuitable. It was rare for anyone to be declared "inkompetent". However, when it had happened in the past, it was only by one of the three or four members of that specific committee who used this weapon.[106] The committee deciding about the chair in Uppsala had not only put him down; they had drawn a bottom line – what Dovring did was not, and should not be regarded as, proper historical research.

Dovring's hierarchical position was weak: he had published at a pace surpassing most other candidates, which endangered the plans for the distribution of chairs made by the leaders of both groups, and finally he had attacked Lönnroth. Yet this taken all together does not explain the aim to crush Dovring. One must try to find the ideological roots for this decision to destroy his career.

The only evaluator, who declared Dovring as qualified, but with reservations,

105 DA, Excerpts of the minutes of Humanistiska sektionen in Uppsala 1.10.1952/ utdrag av Humanistiska sektionens protokoll den 1 oktober 1952.

106 Gunneriusson (2002), p. 90–91, 93 about the declaration of Kumlien, (kk) in the sociogram, as "inkompetent", also in this case very much because he had questioned Lönnroth's interpretations.

was Sture Bolin, who felt some responsibility for his former student. Bolin remarked that he did not believe in some of Dovring's results, and that Dovring had not proved his capacity in the fields where most historians work. Bolin then conceded that: "no other Swedish historian has worked with non-Scandinavian fields of research as profoundly as Dovring."[107] This is a recognition of what many must have thought about Dovring at this time. In spite of this, Bolin ranked Dovring last among the candidates.

The Norwegian professor Sverre Steen questioned why Dovring did not take Norwegian agrarian historians into consideration.[108] This was a correct remark. Dovring had overlooked important Norwegian research in agrarian history. He ought to have quoted them, especially as they worked with retrogressive methods as did he. His whole life Dovring would be fastidious in quoting others, which made those overlooked and their friends inimical to him. That a leading Norwegian historian was one of the evaluators did not work in favor of Dovring.

Nonetheless, Steen admitted that Dovring had considerably expanded the knowledge about land measurement, and that he had conducted his methods with care. When Steen declared him unqualified it was mainly because Dovring had worked with agrarian history and not with general history.[109] This probably did not imply that Dovring's research experience was not broad enough, but rather that it was not applicable to this appointment.

Curt Weibull, the younger of the Weibull brothers, made the third evaluation. He was known to have been outraged when his disciple Erik Lönnroth was attacked. His evaluation was a character assassination of Dovring. According to Weibull, the contribution of Dovring was mainly to give further support to already established truths by collecting material, and Dovring's own interpretations were of no value.[110] This is a variant of the saying: what is good is not new, and what is new is not good. Weibull went on to state that Dovring's first book, *Attungen and marklandet* could not have made him a "docent" in history, only in the limited subject of agrarian history. A large part of Weibull's evaluation was a detailed defense of Erik Lönnroth's interpretations, and he maintained that the results in Dovring's thesis were ill-founded in the sources.[111]

In commenting on the book about taxes, Weibull explained that Dovring had severe limitations as a historian, as he did not connect his study about taxes with "politics in general, and especially the military circumstances".[112] Weibull here gave a declaration as to what should be considered as history proper. With those limits most historians today in Sweden would have been excluded from the craft.

107 Bolin (1952), p. 32.
108 Steen (1952), p. 18.
109 Steen (1952), p. 22.
110 Weibull (1951–52), p. 218.
111 Weibull (1951–52), p. 219.
112 Weibull (1951–52), p. 220.

Weibull concluded that Dovring's book about taxes was: "not remarkable as a scientific achievement of higher quality" and it was comparable to a thesis below the average (in Swedish "en ordinär 2-betygsavhandling"). This is probably one of the worst underestimations made in Swedish historiography.[113]

Curt Weibull's conclusion was that Dovring actually lacked capacity to be a scholar: Dovring had by sheer luck made some finds in the archives (as the notations by Hugo Grotius), and had made some smaller contributions to agrarian history in Europe. Indeed Dovring's lack of skill in the craft of history was said to be the main reason why he could not be considered as competent to hold a chair in history.[114]

The question about Dovring's competence was then discussed by the faculty. In the minutes the professors were divided. Some declared that it was unreasonable to declare Dovring unqualified because he had worked with agrarian history; others supported the majority opinion of the committee.

Most elaborate in his comments was the faculty member and professor in history Erik Lönnroth. He was as impudent as Curt Weibull, and he, without reservation, agreed with his former teacher.[115] As the skilled literary stylist that he was, Lönnroth played with formulations to humiliate Dovring. He said to the protocol: "Dovring's research is characterized by a sympathetic ambition to shape his own opinions and proclaim them with vigor, the first mentioned does not succeed as often as the last mentioned."[116] Lönnroth concluded his comment with the statement that Dovring has "so many peculiar and not well thought-out ideas" that it would be impossible for him to have a chair in history. Lönnroth then makes the following prophetic statement: "lately he has started to do non-Nordic agrarian history, but I must question if his research method will be better if exported."[117] Again this is one of the more astonishing comments made in Swedish historiography as Dovring actually later forged an international career, using his method.

A vote was taken in the faculty whether Dovring should be considered as qualified or not. All proceedings were held in public. In early September 1952, in the Uppsala faculty of liberal arts, 18 voted for his competence, many of them declaring their hesitation, and 9 voted that Dovring was unqualified to hold a chair in history. The result of the vote was published in the local paper in the university-town of Uppsala. This was an ongoing public humiliation of Dovring.

113 Weibull (1951–52), p. 220.

114 Weibull (1951–52), p. 221.

115 Excerpts of the minutes of Humanistiska sektionen in Uppsala 1.10.1952, p. 31, the Folke Dovring archive, utdrag av Humanistiska sektionens protokoll.

116 Excerpts of the minutes of Humanistiska sektionen in Uppsala 1.10.1952, p. 32, the Folke Dovring archive, utdrag av Humanistiska sektionens protokoll.

117 Excerpts of the minutes of Humanistiska sektionen in Uppsala 1.10.1952, p. 32, the Folke Dovring archive, utdrag av Humanistiska sektionens protokoll.

Indeed today his relatives in Uppsala still remember that they discussed and wondered why he was treated this way.[118] Dovring had also applied for the chair in Gothenburg, in 1953, and was again ranked as last. Furthermore his incompetence was mentioned by Curt Weibull, who sat on the committee.[119] In the competition for the chair in Uppsala, in 1954, Dovring withdrew his application in September 1954.

As Dovring's major works were never reviewed in journals by Swedish historians, these evaluations are all the more important in understanding why he was excluded. It is especially significant that the comments were made by influential professors, such as Curt Weibull and Erik Lönnroth. Two general assertions are obvious. One is that the history of the State was considered as history par préférence. To study how taxes fell upon peasants was something other than real history. Agrarian history, as such, was simply not considered as history proper. The second general contention concerned methods. Dovring's statistical method was considered as unreliable and odd. In both these concepts, the leading Swedish historians were behind the frontier of European research of the 1950s.

And, not to be forgotten, his attack on Lönnroth was an offense to be punished. Still, in historiography ideas count and it was history from below and numerical methods that offended most.[120] This battle shows us the basic consensus among Swedish historians of that time, that otherwise seldom was revealed.

The battle phase 2 – fighting retreat

Folke Dovring did not have many adherents. He got some supportive letters from archivists he had met during his research and from others he had known for a long time. The support from the "conservatives" was with one exception non-existent or weak.

A most interesting reaction came from Sven Ulric Palme, a close friend of Erik Lönnroth. Palme was also marginalized in the competition. In a first letter to Dovring in the summer of 1951, he talks about his and Dovring's common unhappy fate. Even though he did not know Dovring before, he advises him not to make a complaint. He also suggests, jokingly, that Dovring perhaps in the future could get a "peasant-professorship" (Swedish "bonnforskning").[121] It is not im-

118 Interview with Dovring's niece Maj Sundblad, 3.30.2002.

119 DA, Expert opinion concerning the chair at the faculty of history in Gothenburg 1953/ sakkunnigutlåtande Göteborgs högskola.

120 Gunneriusson (2002), normally chooses explanations from academic sociology when he discuss the internal strife in Sweden, but in the Dovring-case, p. 160, he declares that the controversy was about what should be counted as history.

121 DA, Letter from Sven Ulric Palme to Folke Dovring, 21.6.1952. And an undated letter commenting on the article in the local Uppsala newspaper, congratulating Dovring as he was not totally cast out.

possible that this was a reproduction of what Lönnroth earlier had suggested about a teachers position in agrarian history at the College of Agriculture.

Palme and Lönnroth took part in a coffee-drinking group at one of the old coffee-shops at the university at Uppsala, together with members of the faculty. Some of these members were the same who later voted that Dovring was unqualified, such as the professor in philosophy Konrad Marc-Wogau.[122] Perhaps the Dovring-case was discussed in their informal group, and Palme's letters can have been a reflection of these chats.

Even though Palme later turned against Dovring, he and his fellows, at this stage, were perhaps prepared to forgive Dovring had he just been a little more meek and respectful to the principle figures after he had gotten his beating.

But instead of keeping a low profile and waiting for a highly uncertain professorship in agrarian history, Dovring decided to strike back and wrote an apply pamphlet (as did other candidates). When Palme later read this in the summer of 1953, he sent a short note to Dovring where he declared that he, Folke Dovring, probably had weakened his own cause by writing a complaint. It was especially detrimental that Dovring had tried to attack and criticize the other candidates as well.

Dovring spread his pamphlet quite widely. One line of defense was that the evaluation committee had not taken his international reputation into account. Dovring also commented on the evaluation committee's disregard for agrarian history, which he felt was rooted in contempt for peasant society in general. Another subject he lingered on, was that he was punished for having attacked the Weibulls, especially Lönnroth. Dovring declared that the Weibulls had turned into "a school of thought becoming rigid into conservatism in relation to their own results", meaning that they wanted to keep their own truths.[123]

He stated that many historians built their hypotheses on blurred and vague sources, and he then proceeded to declare that therefore: "many of the other candidates' publications are, from a fundamental standpoint, inferior to my publications", and he felt quite alone among Swedish historians with his reliance on mass data. Dovring demanded to be ranked among the three top candidates.[124] Such a demand is proper in every apply pamphlet, but the way Dovring lumped everyone else together was probably not very tactical. This did not make him popular among his fellow historians. He also launched a specific attack on his supervisor's favorite, Sven A. Nilsson. Dovring proved that he, in the same period, had been much more productive in publishing than Nilsson.[125]

The reaction came in a published pamphlet from Nilsson, where he stated that

122 Gunneriusson (2002), p. 165, 180.

123 Dovring (1952), p. 6.

124 Dovring (1952), p. 18.

125 Dovring (1952), p. 19–20.

one must trust the evaluators – Dovring was incompetent.[126] Such strong language, however, was the style usually applied in these pamphlets. Probably the lifelong hatred that the publishing of them caused was one of the main reasons why this specific academic tradition succumbed.

Although rejected by many historians, Dovring received support from people he had not known before. In late 1952 he got a letter from Adolf Schück, who was responsible for a series of popular publications connected with the journal *Historisk tidskrift*. Schück was a librarian in a scientific library, and affiliated with the "conservative" group of historians. He began his letter by praising Dovring for the apply pamphlet, and then he invited Dovring to write a popular survey about agrarian history. The proposal was that Dovring should defend this field of research as an answer to his opponents.[127]

The book, *Agrarhistoria* ("Agrarian history"), which I will treat more extensively below, did not at all attack his enemies. Instead he tried to give a positive program for a future new subject.

Later he was invited to write articles about taxes in the huge lexicon project about the Middle Ages in the Nordic countries, *Kulturhistoriskt lexikon för nordisk medeltid*. Dovring also got the chance to talk on radio about his new project: the modern history of European agriculture. So he was not left totally of the arena.

Dovring eventually came to identify Lönnroth as his main adversary. Before he left Sweden, Dovring wrote a long article about Lönnroth's major work about power and finances in the state, *Statsmakt och statsfinans* ("State power and finance"), published in 1940. Dovring's article was published in 1954 in the leading journal of Swedish history, *Historisk tidskrift* (actually it was the 1953 number of the journal, but it was not printed and distributed until 1954), which was mainly in the hands of the "conservative" group of historians. They handled the article as a hot potato.

The article had an introductory footnote, which was a unique comment from the "editors" who declared that they had been hesitant to publish it, but several reasons had been taken into consideration, not the least of which was that the author, Dovring, had an international reputation.[128] Probably they were afraid that he would publish it abroad, had they not published it.

The article caused a crisis in the history of this venerable journal, and the Weibulls afterwards forced through a new organization for the journal, in which they had more control. An elected board was established, which would authorize the publishing of articles. Lönnroth was one of the members, and the Weibulls had the majority.[129]

Dovring explained in his article why he wanted to review this book by point-

126 Nilsson (1952), p. 4.

127 DA, Adolf Schück to Folke Dovring, 22.9.1952.

128 Dovring (1953), p. 384.

129 Schück (1980), p. 130–132.

ing out that he had not before realized the extent of Lönnroth's shortcomings. In fact Lönnroth's poor handling of sources, and his conclusions based on very scant material made the book totally useless.[130] And, as the book never had been reviewed and was considered by historians to be important, Dovring wanted to criticize it. He starts with a defense of agricultural and agrarian history. Then he comments on one of the more remarkable statements made by Lönnroth; that conflicting interests between the landed nobility on the one hand and the peasants on the other hand did not exist during the Middle Ages in Sweden. Dovring considers this as not being supported by the sources.[131]

Folke Dovring quite extensively repeats his earlier criticism of how Lönnroth interpreted the introduction of taxes and rents during the Middle Ages and enlarges it based on his European research about large estates in the early Middle Ages. Lönnroth held the opinion that Sweden, apart from nearly all other countries in Europe, never had a system of large estates. Being much more of a European scholar, Dovring stated that Sweden had about the same development of the social structure as much of the rest of Europe. Based on his studies in Sweden, France, Switzerland and Germany, he talked about a parallel change of organization and administration. Large estates, often cultivated by slaves, where divided into smaller farms during the centuries around AD 1000. At the same time as Dovring presented his hypothesis, French scholars proved the same, and later even Swedish research has agreed with Dovring's standpoint. Lönnroth's position has today become obsolete.

However, to a large extent, Dovring devoted his article to details. For instance Lönnroth had not gone through unpublished sources, and thus had missed that the fall in prices of land had already started by 1360 and not in 1400. This obsession with details could seem to be a little bit fussy, but for Dovring this was probably the most important part of the article – it was about the craftsmanship of a historian. Dovring concludes that a synthesis must build on decades of hard scientific research, and he gives as an example Marc Bloch's *Les caractères originaux de l'histoire rurale française*, thereby again pointing to the *Annales*-tradition as the prototype.[132]

Dovring had slammed the door behind him before he left.

The battle phase 3 – a new program for agrarian history

Dovring wrote two general textbooks about historical research, published in 1953 and 1960. One of them was a research program for agrarian history, as it would have been formed had he stayed in Sweden. The other was a firm statement about methods, a credo to facts replacing speculation, which I will present later on.

130 Dovring (1953), p. 407.

131 Dovring (1953), p. 404.

132 Dovring (1953), p. 409.

In the book on agrarian history (*Agrarhistoria*) a different author appears. He writes a clear and fluent prose; he paints large historical sequences with a few sweeps from his pencil. This is not because he had left his earlier obsession with details but because he is talking to a general audience. One also gets the impression that Folke Dovring knew that he, for a long time, would not turn back to history and he was prompted to posit what he wanted said about the subject.

The book published in 1953, was partly an answer to his antagonists, but he did not mention this and, actually, the text goes much further. He sketches a wide program for historical research, touching upon a range of different subjects where he gives his opinion, which often is both original and sound. In much of the work, he is in advance of research carried on later. Indeed, if at least essential parts of his research program had been carried out in the 1950s, Sweden would probably have enjoyed a more advanced position in the historical science of Europe.

He starts to contest the idea of agriculture and the countryside as something backward: a constraint to change.[133] The town and the higher classes have often been seen as triggers of change. This is a misconception. Agrarian change is of fundamental importance in pre-industrial societies. In addition, if there are elements of conservatism and continuity in the agrarian society, that is because there is a reason. In dealing with small margins, when the harvest barely covers the need and often not, experiments can be dangerous. Biological rhythm as such also gives a certain and much needed continuity to agriculture throughout history.[134]

Dovring does not however underestimate the role of the landlords, and talks about how the rulers of a society mark the whole culture and social structure.[135] In this balanced interpretation of a new role for the peasants in history, Dovring was in line with the frontier of research in the world (and far ahead of the Swedish historians at that time).

He explains that, even though agricultural history in Sweden is looked upon as an oddity, it is different in other countries. He describes Marc Bloch as being foremost among agricultural historians, but also mentions others in the French tradition. He requests a similar development in Sweden, alluding to what a well-known Swedish author (Vilhelm Moberg, who wrote the great epos about the emigration from Sweden to America) had said: the history of cultivating the fields was the backbone of the history.[136]

After this pledge for agricultural history, he goes on to describe settlements and population, cultivation and productivity, land area measurements and land-ownership, taxes, classes in the countryside, agrarian conservatism and peasant revolts. The book is so rich in ideas that it is impossible to mention them all.

He also describes his *kasuistik*-method, but in this book he takes it one step

133 Dovring (1953b), p. 4.

134 Dovring (1953b), p. 6–7.

135 Dovring (1953b), p. 5.

136 Dovring (1953b), p. 14.

further and talks about evidence out in the fields as a complementary source besides maps and cadastres.[137] He describes traces of large fencing systems in Östergötland and other provinces, and even presents maps of old remnants of fences. In accordance with adjoining grave mounds, he suggests a dating to the first part of the first millennium AD. Here he points at a source which was to become dominant in historical geography from the late 1970s, twenty years later. Dovring was correct in his estimation of the dating, and in his identification of these stone heaps as remnants of fences.

He emphasizes the role of the village maps from the seventeenth century as a source for agrarian and agricultural history.[138] These maps are today one of the most utilized sources, especially for landscape reconstructions. Dovring also mentions accounts from sixteenth century royal farms as an underestimated source.[139] As I have worked with that source, I can only agree. These accounts are probably among the best from Early Modern Europe, but are still an underused source.

He summarizes the research on land area measures in Europe and declares landholding in Sweden to have been a part of a European system prior to AD 1000, with large units and un-free labor.[140] He here confronts the historians of his own time but is in line with what historians think today. Today most historians would accept that the "free farmer" in prehistoric Scandinavia belongs to the myths of history. Instead hierarchical structures have a long history. He presented the same ideas as he had in the article published in *Historisk tidskrift* the same year, 1953, where he also pointed to French research as his source of inspiration, but now it was meant for the general audience.

In the book he also depicts the historical development of settlement in large parts of Europe, thereby pointing out the direction his research was taking at the moment. He also makes bold connections. For instance, he depicts the farm structure in North America as influenced by ideals governing nineteenth century North-western Europe, with "independent yeoman farms". Latin America's system was, in contrast, formed by an earlier epoch and another part of Europe, the Iberian large estate with its concentrated settlement.[141]

For demographical research, he suggests investigations of the number of children in different social strata in nineteenth century Sweden.[142] The hypothesis is that the peasants had some means to control the number of children – which would imply that different strategies could be developed among different groups of the rural population. Exactly such research was successfully performed in Sweden in the late 1970s. All together his book was quite advanced for its own time.

137 Dovring (1953b), p. 27.
138 Dovring (1953b), p. 58.
139 Dovring (1953b), p. 29.
140 Dovring (1953b), p. 99.
141 Dovring (1953b), p. 18.
142 Dovring (1953b), p. 113.

Responses to the challenge

Lönnroth never answered directly in print to Dovring's attack, and never published a reply in *Historisk tidskrift*. He seldom actually mentioned Dovring. Lönnroth later emphasized more fully the importance of international contacts, and in 1963 he became one of the few Swedish historians that published a chapter in the prestigious *Cambridge Economic History of Europe*. Sture Bolin had published in the first volume, mentioned in a foregoing chapter, and in the 1960s Folke Dovring published a chapter (see below). During the following decades a couple of economic-historians made contributions, but very specifically about Sweden.

Lönnroth's chapter was about government finance and trade in the Baltic region during the Middle Ages, wherein he persisted in his interpretation; taxes paid to the government formed the new land measuring system. "The new system of taxation had to be based on individuals, each economically capable of being regarded as a unit for purposes of paying taxes"[143] Dovring's book *Attungen and marklandet*, where he criticized this theory of Lönnroth, was not even mentioned among the some fifty different books Lönnroth has in his list of literature contributing to the article, but Dovring's tax-book, where he did not criticize Lönnroth was in the list. Lönnroth had little interest in considering a revision or discussion of the results he once had laid fast with his "exact and definite" source-critical method.

Dovring's programmatic book *Agrarhistoria* ("Agrarian history") from 1953 got a mixed reception. Late in 1953 Sven Ulric Palme, Lönnroth's friend in Uppsala, wrote a rather acid review in the daily newspaper *Stockholmstidningen*.[144] He called it a stimulating but chaotic little leaflet. Dovring was said to be hampered by being a fanatic, both in his fight for agrarian history and in his criticism of certain other scholars. Palme explained that historians are concerned with change, and, as agriculture did not change much before the eighteenth century, it is logical that they who study earlier history have not coped with the history of agriculture. On the other hand, considerable contributions in fields bordering on agricultural history have been made by historians. Palme thought that Dovring concealed this fact because he considered these persons as his personal enemies. Palme concluded his little review by blaming the editors, and gave a suggestion that the editor for these publications (Adolf Schück) ought to be put under better control (as had happened with *Historisk tidskrift*). Palme's short review has been given lengthy attention here, because it probably corresponds quite well with the opinions of the leading group of historians – Palme was just more outspoken. Their opinion was that agrarian and agricultural history was not of interest for older periods, and that Dovring had unjustly attacked leading historians (namely Lönnroth). He was a fanatic, and therefore unreliable; his publications ought to be suppressed.

143 Lönnroth (1963), p. 371.

144 Palme (1953).

Two years later Gustaf Utterström, who was to become the leading agrarian historian after Dovring in Sweden, published a review in *Historisk tidskrift*. As an agrarian historian he was mainly positive, but had some objections. He complained about Dovring's insufficient quoting of other scholars; as an economic historian he emphasized this literature.[145] He mentioned that Dovring pointed at environmental causes for the agricultural depression of the Late Middle Ages, which at the time was becoming perhaps the most discussed of all historical questions in Europe.

In Germany the book *Agrarhistoria* was mentioned in a survey article about agrarian history in the Nordic countries, written by Karl Whürer 1957 in *Zeitschrift für Agrargeschichte*. The author compared Dovring with the Norwegian group of agrarian historians, who used the retrogressive method in a similar way to Dovring.[146] Whürer also mentioned the controversy between Dovring and Lönnroth, which apparently was well-known outside Sweden. But Whürer did not take partiality with any side in the issue, instead he remarked that a core of the conflict was Dovring's early dating of large estates.

After the battle

In the long run, the approach Dovring championed came to be victorious also in Sweden. Numerical analysis was the upcoming method. Especially concerning later periods, numerical analysis came to dominate totally. Early examples are the New Modern period with massive works by Birgitta Odén (who was a friend of Karin Dovring) and, for later periods, the analysis of social groups made by Sten Carlsson.[147]

Agrarian history suffered a temporary setback, but soon Sweden started to catch up with the international trend. In 1956 Dovring's fellow graduate student Sten Carlsson, published the third volume in the series *Bonden i svensk historia* ("The farmer in Swedish history").[148] Lönnroth and his allies were not amused, and in letters they scorned Carlsson for working with peasants, and considered it merely a way to flatter the politicians in the farmers' party.[149] An important contribution was published by the economic historian Gustaf Utterström in 1957, about the agricultural change during the eighteenth and nineteenth centuries.[150] Agrarian history in Sweden had to wait, however, until the 1970s for a more gen-

145 Utterström (1955), p. 331.

146 Whürer (1957), p. 200.

147 Gunneriusson (2002), p. 171 writes that Sten Carlsson, who published his thesis in 1949, was the first to extensively use statistical methods. But Dovring had accomplished that two years before.

148 Carlsson (1956).

149 Gunneriusson (2002), p. 188, 191.

150 Utterström (1957).

eral progress of the subject. Then a number of quite important works were published.

Lönnroth's dominating influence did not only affect Dovring. Another example can be given from a large Nordic project during the late 1960s and the 1970s on the crisis around 1350–1500. This was a project in medieval agrarian history, which Erik Lönnroth initiated and for which he found funding. Some of the methods Dovring had suggested were used, but the Swedish part of the project was hampered by the urge to defend a thesis Lönnroth had previously put forward in the 1930s. He had assumed that large parts of Sweden, not least the woodlands, went through a period of expansion and clearances during the Late Middle Ages. This was contrary to what European research had shown from the 1940s: namely a tremendous decrease in population and settlement all over Western Europe the decrease being most severe in the woodlands.

Lönnroth seldom gave up an idea and, even though some young Swedish historians in the late 1950s had proved that Sweden went through a crisis as detrimental as most other nations, their results were for a time pushed aside. Lönnroth held on to his idea of Swedish peculiarity. In his 1963 article in *The Economic History of Europe* he wrote on the Late Middle Ages "the agrarian recession of the period seriously affected Denmark and Norway and to a lesser extent Sweden".[151]

Lönnroth's old hypothesis lay as a drag on the project. Accordingly, the Swedish scholars developed methods to get the portion of deserted farms as low as possible. Indeed the Swedish and the Norwegian parts of the project developed totally different methods. Therefore, the map of Late Medieval desertion, presented by the project, mainly showed the distribution of different research methods in the 1970s. It must be said that several of the investigations made in the Swedish part of the project evidenced held high quality research. Notably, some of these historians even expressed an implicit criticism of the chosen methods. This eventually, but much later, led to an acceptance of a more reasonable picture: Sweden was hit by the plague and the ensuing recession as strongly as the rest of Europe.[152]

It must also be stated that others of the Weibullians were more open-minded than Lönnroth, as was, for instance, Dovring's old competitor Sven A. Nilsson. After he had gotten the chair Dovring applied for in Uppsala, and not the least during the student-revolution in the 1960s and 1970s, Nilsson managed to attract some of the most talented of these students. He took them as pupils even though they held radically different views than those he himself advocated.

It takes more than one man to shape a milieu. A most interesting question to pose for further research on Swedish historiography is to ask: how these dysfunctional elements developed that allowed the leaders' unjust attack on Dovring?

151 Lönnroth (1963), p. 381.

152 Myrdal (2003).

Perhaps some understanding can be achieved from a wider perspective on Swedish universities in the post-war period.

Conformism around 1950?

Folke Dovring was far from alone in being pushed aside in the academic society in the late 1940s and early 1950s. There will always be disposal of individuals from academic society, persons misunderstood and underrated in their own time but rehabilitated by posterity. Scientific research must operate that way. New ideas and theories go through the mill of doubt and examination.

The question is wether Dovring were excluded during a period marked by strong conformism or not. I will give a few examples, but I do not intend to give a total exposé of remarkable bans laid on scholars during different periods. For a more thorough study one must also perform a comparison with conformist scholars and their relation to the reformists.

Gunnar Ahlström (1906–1982) had presented his thesis in 1936, in Swedish literature. He became docent the year after, and, thereafter, published books on Selma Lagerlöf and other Swedish authors. In 1947 he published what was to become his most important work. The book gave a broad description of the breakthrough of modern prose in the Nordic countries during the late nineteenth century. His interpretation was Marxist inspired. He pointed to the change of fundamental social structures, shaping a new middle class to be portrayed and at the same time forming a new market for literature. Social and economic change led to great literature like works by Ibsen and Strindberg. Ahlström was immediately attacked in the daily press as unscientific. In the following year, 1948, in a competition over a professorate, Ahlström was abolished from the academic society. The question of historical materialism was avoided. He was considered as unscholarly. The word "fanatic" was mentioned. He was nearly declared incompetent, and only saved by his long previous career where he had established a reputation as a conventional scholar. Ahlström had to leave the academic world and later became an administrator at The Swedish Institute (Svenska institutet, SI), a public agency that promoted Swedish interests abroad. After spending the years 1949–1955 as the director of the section in London, he returned to Sweden where he worked for the institute until he retired.[153] His book was republished in 1973 and today it is considered as a standard work in sociology of literature, a specialty which has since been established as a strong field of research in Sweden.

Another example is that of Börje Hanssen (1917–1979). He started as an economic historian, then changed to sociology, and tried to use concepts and theories from sociology while investigating rural society in Southern Sweden during the seventeenth and eighteenth centuries. His dissertation of 1952 was ques-

153 Forser (1978), p. 145–150.

tioned, and when he tried to get a docent-grade, the only "ticket" to a further academic career, he was nearly refused. Only after a long and dirty conflict was he given a docent-grade in "historical sociology" instead of "sociology". Thereafter he was declared incompetent when he applied for professorships in sociology in 1954 and in ethnology in 1972. He corresponded with Dovring in 1953–1954. Dovring told him that he should not be too concerned with the labeling of the docent-grade. Hanssen considered Dovring's emphasis on quantitative data as fair strike against the established historians. Hanssen became a close friend of Robert Redfield, and Eric Wolf thanked Hanssen in his acknowledgments in the classic *Peasants* of 1966. Having many children, Hanssen (though he worked in the U.S.A. 1957–1958) did not want to leave Sweden. Instead he tried to find another opening in Sweden. He lectured in ethnology in Stockholm in the early 1960s, but a conflict with the professor in ethnology there, who declared that Hanssen's approach was not desired, caused Hanssen to leave the academy in 1963. For several years he made a living as a poultry-farmer in the middle of Sweden. He had grown up as the son of a farmer, and now he tried to implement new methods of farming he had learned during his year in the U.S.A. In the mid 1970s, he was again invited to lecture in ethnology. His dissertation was republished in 1979. He then inspired a whole generation of students, but unfortunately died prematurely.[154]

A third example will be taken from natural science. Georg Borgström (1912–1990) presented his thesis in plant physiology in 1939, became docent in 1940, but left the university after a conflict with his professor. He soon was able to establish a strong position in the society of scientists as the director for an institute of plant research in 1941. Later, in 1948, he became the director of the Swedish Institute for Food Preservation Research. He had fought for what he considered to be righteous values since he was a student. For instance, as a leader of the student union, he supported Jewish refugees in 1937–1938. Around 1950 he became more and more engaged in environmental questions, and started to publish and lecture widely about threats to the planet. This led him into conflict with the stakeholders from the packing industry and other industrialists, which had a strong position in the institute. In the autumn of 1955 he was dismissed by the board of the institute, and, after a long and public debate, he left in 1956. He went to the U.S.A. where he immediately became full professor in food science at Michigan State University and an American citizen in 1962. He continued his campaign and became a well-known debater in the U.S., connecting to other conservationists in the country.[155] Basically he was a Malthusian. Some of his warnings have not yet come true, while others, like his prediction of depletion of

154 Perlinge *Manuscript* to a biography on Hanssen, which Anders Perlinge has been kind enough to let me read and quote.

155 Linnér (1998) and Linnér (2003) has in detail described the Borgström-case and the connections with American conservation ideology and international environmentalism.

fish in large parts of the ocean, have unfortunately materialized. In Sweden Georg Borgström's books had an enormous influence on the environmental movement in the 1970s, and he is still discussed as one of the icons of the environmentalists.

As an addendum to these examples of failures, my last example will be a success story. Eli Heckscher (1879–1952) is the most internationally successful historian Sweden has produced. He always wandered along the border between economy and history, starting as a historian but later working in the most recognized business school of Sweden, Handelshögskolan (Stockholm School of Economics). His first proposal on the necessity of economic history was advanced early on, in 1904, and eventually, after many years of arguing, creating intrigue and not to forget publishing copiously, he managed to formally establish the subject of economic history in Sweden in 1948.[156] Economic history later came to be a most successful field, taking the lead in much of historical studies in the 1970s and later. The important difference between this and foregoing examples is that the success of economic history had its roots in the 1920s and 1930s. Eli Heckscher was awarded a personal chair in economic history in 1929, after having been professor in economics since 1909. He had for a long time belonged to the establishment. As a matter of fact, he took part in the attacks on Ahlström, declaring that he did not understand economic history. Heckscher gave support to Hanssen, and also favorably estimated Dovring, but it seems to have been too late in Heckscher's life to help them out of their difficulties.

These examples all involve scholars crossing established borders between disciples. Dovring as an agrarian historian was excluded by historians of the time as his field was defined as not belonging to history. Ahlström introduced economic and social history into the field of literature, besides being inspired by Marxism. Hanssen proposed that qualitative data, necessary when working with historical sources, be included in sociology. Even the response to Borgström can be understood this way, when he started to combine natural science and social science. It is interesting to note that three of them, Dovring, Hanssen, and Borgström, traveled to the United States in the mid and late 1950s.

Heckscher established a new discipline, but this was accomplished basically in the interwar period, and then both with the support from the working class parties and from industry. The goal was to write the history of a new industrialized society. His success can thus sustain that there were different periods in academic excellence criteria, where some were more open to new initiatives.

In Swedish historiography the theoretical concept "doxa", which Pierre Bourdieu introduced (in a description of a peasant society but which since then has wider applications), is often used. Bourdieu wanted to find a terminology to describe how an established order is upheld, where an arbitrary limit is made the

156 Odén (1991) p. 80–86, Hasselberg (2007) on the network Heckscher built during the 1920s and 1930s to reach his aim.

natural limit of thought and behavior. As long as no one questions these basic opinions, the doxa will be upheld. However, when critique brings the undiscussed into discussion it may could lead to a crisis and a change of consensus.[157] This does not explain why an established consensus of arbitrary limits in social science and humanities was stronger in certain periods, and why they seem to have been upheld longer after World War II in Sweden than in the U.S.A.

The socio-political structure in society as a whole could have played some role. During the establishment of The Cold War, McCarthyism in the U.S.A. had its equivalents in many other countries in the West (and probably also in the East). A stronger political contradiction in the world spilled over to the universities, for a while. This has been suggested as a major cause for the exclusion of Dovring, Hanssen and others.[158] However Ahlström was only partly criticized because he was Marxist-influenced. The main attack was on his attempt to develop a social history of literature. Dovring and Hanssen on the other hand were both definitely not Marxists.

They actually did not have a common agenda. Dovring argued in favor of quantitative methods against the historians, Hanssen in favor of qualitative methods against the sociologists. Dovring and Hanssen both had a bottom up approach. That did exclude Dovring, but Hanssen worked among the ethnologist who hailed the peasant society.

Academic sociology and networks helps some way to explain their common failure, as both Dovring and Hanssen were aggressive loners, but many other persons with that personality succeed in the academia. To explain theses examples with common causes one instead has to discuss the university structure.

A general explanation for this unwillingness of the established disciplines to accept scholars who did not want to be restricted by accepted consensus on borders between disciplines could be that Swedish universities had a hierarchical structure. Particularly, one where a few professors defined what was to be accepted. Thus ideological structures constructed before the World War II could be preserved for a long time. Scholars lingered on with discussions becoming more and more out of tune with the international intellectual discussion.

A strong inbreeding characterized, and still characterizes, many fields of research in Sweden. This is a constant risk in a small country. In some fields of research, a person can spend his whole career, from student to professor, in one department, and actually in one hallway. This does not work in favor of unconformistic behavior.

A third structural reason could be that only a small portion of the population attended universities up to the 1950s. This was later to change. In the late 1960s and in the 1970s the number of students and also the number of associate profes-

157 Bourdieu (1977), p. 164–168.

158 Nilsson (2004), p. 131–134, who maintains that socio-political factors were decisive but he does not analyze the difference between actors in the drama.

sors (lektorer) and full professors greatly increased. Many of the former dictatorial professors had to relinquish much of their power.

Sven-Eric Liedman, a professor in history of ideas, has in several books described these conservative structures, which dominated until the 1970s. Not unexpectedly Erik Lönnroth, who at that time had established himself as the most powerful figure in the whole field of humanities and social sciences, attacked Liedman fiercely. But that is another story to be told, from a later period than the one I am discussing here.

If different periods can be identified, in the Swedish academic society, as being more open-minded (1920–1930s), more narrow-minded (1950s), and again perhaps more approachable (1970s), this pattern can be mainly explained by changes society's ideology and social structure. However, the specific reason for the late 1940s and 1950s as representing a period when the academic society tended to expel contenders was probably the lingering structural deficits in the Swedish university system. These were retained for a long time after the World War II and tended to give an advantage to the leaders of the hierarchy.

CHAPTER 4
Years of hardship and success (circa 1953–1969)

Wandering years

There are points in life where one has to choose. For Dovring such a juncture came at the end of the long autumn of 1952. The goal he had hitherto sought was blocked. He would never get a chair in Sweden. One opportunity was to continue with European medieval history. But if he tried to go abroad, to Europe, there was a probability that Bolin or others with international contacts could hinder his way. Lönnroth also had such contacts; his comment about the eventual failure of "exporting" Dovring's ideas could be a threat.

Thus during the years 1953–1960, Dovring shifted both his place of residence and his direction of research. His change of attention started when he began to take more interest in European history, even before he left Sweden. In the mid-1950s, he also changed the time period studied, and turned to later agrarian history. From 1960, when he had arrived in the U.S., he started to take up agrarian problems in contemporary society, and also to change focus to countries outside of Europe. After about 1970, he rarely wrote about history and seldom about Europe. This change is illustrated in figure 3.

As early as 1951 Dovring had started to ponder about a change of research to later periods, and applied for money from the Swedish Council for Social Sciences to make a long European odyssey. Dovring was successful in getting the grant and, in October of that year, he started an eight-month tour through Portugal, Spain, Switzerland, Greece, Italy, Yugoslavia, Austria and France. In every country, he spoke with agricultural experts, visited archives and libraries, and traveled to look at the countryside with his own eyes. He wanted to write a book about the modern history of European agriculture, partly as a follow up to his earlier research on the Middle Ages.

He had received the grant just before the big battle started. Perhaps a reason why he got it was that the decision preceded the merging of funds for humanities and social sciences into one large fund-granting council. Therefore his foes among historians, who controlled the Council of Humanities, had no influence over decisions taken in the Council of Social Sciences. When he came back from his long journey, he was already deeply involved in the battle. When it was all over, he was

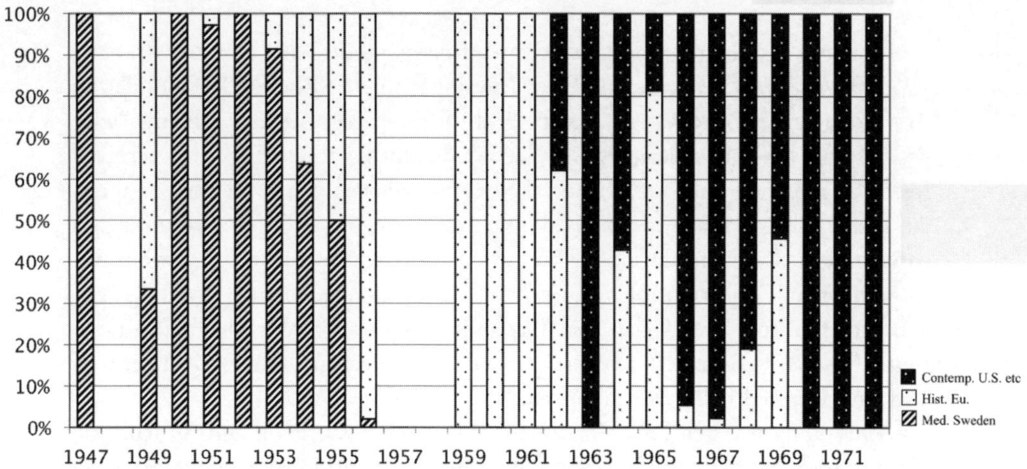

Figure 3. Dovring's publications subdivided in subjects 1947–1972. Subjects: Medieval Sweden; History Europe; Contemporary U.S. & other countries. Second editions not included. Source: Dovring's bibliography, Appendix 1.

an outcast who would have difficulties in finding a position where he could finish the book.

However, he came upon the opportunity to finish the book by a strange coincidence. When I started to do research on Dovring, I had no hint about any connection between him and me, besides that we both were agrarian historians. My grandfather Gunnar Myrdal never mentioned Dovring, though he of course knew that I worked with agrarian history, and I did not ask about Dovring because I had no idea of any acquaintance. Gunnar Myrdal was a famous economist, who wrote about the racial questions in the U.S., about the Indian economy and about many other topics; eventually he was awarded the Nobel Prize in economy.

Gunnar Myrdal was hit by a car in Germany in the fall of 1952, and was transferred to the university hospital in Lund, where he spent the whole November of that year. Few visited him besides his relatives, but one academic who went to see Gunnar Myrdal was Folke Dovring. They had had some contact before, probably when Dovring, in 1951, visited the Economic Commission for Europe (ECE) in Geneva, of which Myrdal was the director. Dovring's fellow at the department of history, Gunnar Westin, when interviewed said that Dovring had reported enthusiastically about his meeting with Myrdal.[159] To Folke Dovring, at that time pressed by his opponents, it must have been a relief to talk with someone who respected him. My guess is that Gunnar Myrdal was not totally unaware of the on-going academic-political battle (he had many friends in Uppsala). For Myrdal it could have been both an occasion to rescue this young gifted scholar and an opportunity to influence him.

159 Interview with Gunnar Westin 3.2.2002.

Probably Myrdal discussed some sort of cooperation with Dovring. In early 1953 Dovring was advised to apply for funding from The Rockefeller Foundation to make a survey of tenure in Europe and to discuss connected problems "with Mr. Myrdal and his colleagues" in Geneva.[160] During the summer, he received from Rockefeller a one-year of "Social Sciences fellowship" in "the field of agrarian history".[161] At that time, he had already been working for two months as a consultant in ECE. In November he started his research year to finish what was to be his major contribution to international research. Gunnar Myrdal had encouraged him to leave Sweden and start an international career. That Karin Dovring, at the same time, left Sweden and started a career in the U.S., at Yale, may also have inspired Folke to try himself.

After doing research for a time in the library of FAO (Food and Agricultural Organization of the United Nations) in Rome, the world center for research on agriculture, he went to Geneva. There he finished his comprehensive study of European agriculture in the first half of the twentieth century, looking back also to the nineteenth century.

In his report to The Rockefeller Foundation, in late 1954, Dovring makes several interesting comments about why he chose to change the subject of his research. (The report is preserved in a draft in the Dovring-archive.) He looked upon his year in Rome, and Geneva at ECE not only as an opportunity to finish his research about recent agrarian history, but also as a training in research on agricultural economics and sociology as part of agrarian history. He wanted to be an agrarian economist besides being a historian. He wrote:[162]

> "when I, some three years ago, decided to finish my research on Medieval and other older agrarian history and take up research on modern problems in the field, it was through a combination of considerations on methodology and such on the purpose of scientific work. Field studies on Medieval villages in France and Switzerland had taught me that much more of the Middle Ages is alive on European countrysides than generally occurs to Scandinavian and English observers."

He also declared that: "purely historical research, without contacts with physical realities as tests on its realities, could to some extent lose the touch with its object." His field studies as a means "of limiting the scope for various hypothetical

160 DA, Letter from Frederic C. Lane to Folke Dovring, 22.4.1953.

161 DA, Letter from The Rockefeller Foundation to Folke Dovring, 24.8.1953.

162 DA, Report November 1 1954 to The Rockefeller Foundation on work under Social Sciences fellowship from November 1952 to October 31 1954. In a letter November 24, recognizing that the foundation had received the report, Marion Elderton asked "Does this temporary work with FAO mean that you have given up the idea of returning to the University of Lund?"

solutions of problems on Medieval villages" had convinced him that current research methods in agrarian history, and in history as such, were in need of reconsideration. But those who were arguing with him had no such experience themselves, and were not convinced of this necessity for a change of methods. Here again, Dovring was hinting about the feud with Lönnroth and other Swedish historians.

He wrote that "as I see it" historical research should be carried out on sources "so recent and so abundant that most methodological developments could be tested". Also, if later on he were to return to "the history of older epochs, I hope that I will from my modern research bring with me such experience on abundant statistical material" which will be helpful for making conclusions on the basis of the scarce material from older epochs. However, Dovring would not return to deep historical research on the Middle Ages. When he had finished his year as a Rockefeller fellow, he obtained a position in Rome at FAO, at first on a short time basis, and then from late 1955 until 1960.

Land and Labor

The book, which was to become Dovring's most important, was published in 1956 with the title *Land and Labor in Europe 1900–1950*, and its third edition was published in 1965. In my description of the book I follow the latest edition.

The first two chapters about village-size, land supply, and labor force were brought together mainly during Dovring's journeys in 1951. Material for later chapters on land tenure and cooperation was collected during his stay in Rome and Geneva.

He considered his book as agrarian history, and, in his introduction, he returned to one of his favorite ideas: how neglected this part of our history has been.[163] But the further the book develops the more the historical explanations fall into the background – even through they are always there.

His method had similarities to the method he had developed in his studies of medieval agrarian history. He collected an enormous amount of statistics. By remolding the statistics step by step and weighing different numbers, he was able to build up a new picture of the situation. The basic data considered village size, area of farms, number of fields and the amount of manpower in agriculture. In particular, he wanted to analyze the relationship between land and labor. Thus the title of the book, *Land and Labor*, describes fairly well his intended goal.

He started with the villages. Dovring constructed a map of Europe showing the size of the median villages in the beginning of the twentieth century.[164] In much of northern and western Europe the single farm or small hamlet dominated

163 Dovring (1965), p. 5.

164 Dovring (1965), p. 15.

with populations of less than a hundred persons. In Central Europe medium-sized villages with an agricultural population of 200–400 were more common. In southern Spain, Italy and also in the eastern Balkans and in southern Russia, very big villages with a thousand inhabitants and agro-towns with several thousand inhabitants dominated.

The dispersed settlements in northernmost Europe were long-established, but, in much of England and France, they were of a later development, with larger villages dominating during earlier historical periods. The very big villages in southern Europe had their roots both in ecological factors, such as the scarcity of water, and also in historical factors. The big agro-town can be followed back to antiquity, but it also reflected the border between the Christian and Muslim regions, where fighting went on for centuries.[165]

From establishing village sizes as a basic factor, he moves on to discuss the layout of land. The fields were divided into small plots, belonging to different farmers in the village. Dovring calculates the average size of the plots, and the number of plots per farm. He states that there is "a tendency to understate the changes going on in traditional rural society".[166] Fragmentation is often described as being characteristic of Medieval traditions, but even though fragmentation happened to some extent in the Middle Ages, much of it was the results of partitions of land due to inheritance over the last generations. Following this line of argument, he also cites his earlier studies about medieval villages in Europe.

His next step is to estimate the distances between the farm and the fields. The average distances were often more than 2 kilometers. Walking to and from distant fields implied a waste of working time, compared with a situation where the fields lay directly around the farm. Dovring concludes that for instance in the agro-towns in southern Europe distances of 5 to 15 kilometers are not uncommon, potentiating a waste of labor. The reason for this permitted waste was the low cost of labor.[167]

Here he arrives at his main thesis throughout the book: overpopulation in the countryside entails waste of labor. To obtain a measurement of this waste of labor, much of the rest of his book about European agriculture in the twentieth century contains a series of calculations about labor force and how it was utilized. In appendices, he presents detailed tables referring to statistics from every country in Europe.

The chain of calculations starts with estimations of total acreage and the number of workers in each country.[168] This includes several tricky estimations comparing statistics from different countries. The male work force, registered by national statistics, is the basis for his calculations. (He makes no efforts to calculate female

165 Dovring (1965), p. 16.
166 Dovring (1965), p. 47.
167 Dovring (1965), p. 30.
168 Dovring (1965), p. 92.

work participation, because of lack of reliable statistics.)

A normal scholar would perhaps have stopped there, but Dovring also tries to calculate the amount of work needed. Efficiency in North American agriculture is used as a yardstick,[169] but also agricultural manuals from different countries. From this he proceeds to make a calculation of the actual amount of available work compared with the need. The difference is identified as underemployment.

He then attacks the problem from another angle. He makes estimations of the average farm size, and, thereafter, calculates the amount of work needed on the average farm, considering both grain and livestock production. He is able to show that the pure family farm dominated.[170] He has thus proved another of his main ideas, that the medium-sized farm (with about 1.5–2 male work-years available) is as effective as the large farm.[171]

At the time when Dovring made his third edition, around 1960, the Soviet system, with collective farms, was an alternative still being discussed, at least in Europe. Using statistics, Dovring is able to demonstrate that the Soviet system not only lags behind the more developed farming of western Europe, it is actually "the most low-productive in the world", with an enormous waste of labor and creativity.[172]

In special sections, Dovring discusses the problem of tenancy. His study also includes a survey of agrarian cooperation. He judges cooperation as necessary since, on the one hand, economies of scale are limited in scope in agriculture and, on the other, a labor surplus situation puts the single farmer in the position of an underdog.

In his description of cooperation, he paints a model, an ideal, which comes very near a Scandinavian farmer's family farm. This is where the one who works the land owns it, and where farmers work together in cooperation when it comes to activities where economics of scales must be used. He would stick to this as an ideal for farming the rest of his life. He sketches a historical background, but mainly he looks upon cooperation as a late development, which came out of a popular movement.[173]

In his discussion about tenancy, he comments on history, which has connections with his earlier research. In the early Middle Ages, there was little fragmentation; large domains were in the hands of the landlords. "Decline of domanial authority in western Europe, mainly in the late Middle Ages, was accompanied by the breaking up of the enlarged family group and the manse."[174] This had been one of the ideas he had propagated in the controversy with the Swedish histori-

169 Dovring (1965), p. 97.

170 Dovring (1965), p. 133–135.

171 Dovring (1965), p. 143.

172 Dovring (1965), p. 108.

173 Dovring (1965), p. 227.

174 Dovring (1965), p. 162.

ans, who believed that Sweden was an exception from general European patterns in this and in many other respects, a belief that Dovring mainly rejected.

His wife, Karin Dovring, wrote a long chapter (actually a quarter of the book) on the ideological basis of different socioeconomic systems. She used a specific method of analyzing what people said and what they meant, implementing a quantitative measure she had developed in her dissertation. This same method opened the opportunity for her to lecture at Yale and other leading intellectual milieus in the world. But I leave her chapter aside.

One of the concluding remarks in Dovring's book is that he had studied the beginning of an ongoing process, and that the number of farmers would continue to decrease. He also remarks that agricultural politics seldom greatly change – a statement remarked upon by several reviewers of the book.

Editions and reviews

Land and Labor was to become the most well known of his larger publications. After its publication in 1956, it was published in new editions in 1959 and 1965. For the third edition of the book, he did wide-ranging research, which became one of his major tasks during his first years in the U.S.

Land and Labor was reviewed in a number of scientific journals. A first wave of about twenty-five reviews came after the first publication. Many of these journals were German and Dutch, but there were also reviews from Italy, France, England, and the United States. In the mid 1960s the third, enlarged edition was reviewed in about ten scientific journals, many of which had already reviewed the first edition.

Most of the reviewers were enthusiastic about the book, and impressed by the amount of material he had collected. In Germany the leading agricultural historian Wilhelm Abel supposed that, for a long time, scholars would go back to Dovring's surveys about village-size, manpower and agricultural land.[175] (Many years later Dovring would respond with a praise of Abel in a review from 1981.[176]) Other commendations were: "a solid contribution" by N.L. Whetten in *American Sociological Review* in 1958, and "to be commended for the wealth of material it contains, for experiments in methodology and, above all, for its suggestiveness of fruitful fields for further research" by Edgard Thomas in the *Journal of Agricultural Economics* 1957. Other words of praise to characterize the book are "fundamental" in *Zeitschrift für Raumforschung und Raumordning* 1957, "remarkable" in *Revue de l'Agriculture* 1956, "original" and "exceptional" in *Economie rurale* 1959. Sometimes the reviews reached high levels of acclaim. Shepard B. Clough wrote in *Agricultural History* 1957:

175 Abel (1957), p. 228.

176 Dovring (1981), p. 296–300.

> “From time to time the scholar finds on his desk a study which is so painstakingly made, which reflects such a complete grasp of the subject treated, and which arrives at such striking conclusions that he could only wish the work were his own. Such is the volume before us.”

Astonishingly, it also was reviewed on August 30th 1956 in the well-known French daily *Le Monde*, where Dovring’s book was presented in an article as “le beau livre”: it has the quality to shape order out of the enormous amount of publications on agriculture.

But there were also sour comments. In *Geographical Journal* 1957 published in London, Harriet Steers complained about all the statistics and mentioned that several English scholars had been forgotten. She concludes that Folke and Karin Dovring “seem to have travelled the length and breadth of Europe with a huge calculating machine between them and the landscape”. In another British journal, *Agriculture,* August 1957, the reviewer N.H. claims that the “faults of this book are obvious and irritating” as the style is dreary, and there are too many local studies mentioned. As was Steers, this reviewer is irritated that some British scholars were omitted from Dovring’s large survey, but also admits that the book “contains a mass of information” and “peculiarly instructive tables”. Interestingly, Dovring did not include the British suggestions for new literature when he reworked the book for the new editions.

When the third edition was reviewed, the book was well known and widely used. For instance, John Marsh wrote in the *Journal of Agricultural Economics* that it was not only a truly scholarly work but also that: “For many students it has already proved its value.” And in the British journal *Agriculture* in 1966, the book was reviewed by J.A.M. who said that: “The general merits of the book hardly need stressing at this stage”, and he also uses the words “extremely interesting and well-documented”. This was a vindication of Dovring compared with the first review in this journal.

Even in Scandinavia (if not in Sweden) his book was reviewed; Viggo Hansen wrote in the Danish *Geografisk tidskrift* 1967 that Dovring’s book was hitherto the best existing survey of agriculture.

By the mid 1960s, *Land and Labor* had been established as a minor classic. For decades everyone who worked with this period of European agriculture considered and quoted Dovring’s book. The quotations in various textbooks up until today are so numerous that it is would be excessive to enumerate them.

Dovring had thus taken revenge, and had proved beyond all doubt, that Lönnroth was wrong when he predicted the eventual failure in exporting Dovring’s methods outside of Sweden.

Attached articles – further success

During his work with *Land and Labor,* Dovring also published some articles about European agriculture, both in Swedish and in several other languages. Among these was one of his most widespread articles: *The share of agriculture in a growing population,* published by FAO 1959.[177] It appeared in English, Spanish, French and Italian, and it was reprinted in English in 1964 and in Spanish in 1968, published in German in 1969, and reprinted a third time in English in 1978.[178] Many economists would particularly hold this as Dovring's most influential shorter contribution.

What seemed to have made the largest impression on other scholars was how he determined the change of the agricultural labor force, with what he termed "the coefficient of differential growth".[179] This measure found its way into standard works and was further developed by other agrarian economists.[180]

Dovring could, with these calculations, prove that "a reduction in the absolute numbers working and living from agriculture is most difficult to achieve when the agricultural population is a large majority".[181] By comparing the share of the agriculture population with the whole population, he also concluded that transition to an industrialized society would be faster the later it occurs. In addition, he recognized that the total amount of people engaged in agriculture, in nearly all countries, tended to be constant for a long time before it plunged downwards. The reason for this degree of stability includes sociological and psychological factors, but also encompasses "expanding markets for agricultural products, and particularly those of labor intensive animal husbandry, and of market gardening".[182] He thus gave instruments for further elaboration of the complicated process in transformation from an agrarian to an industrialized economy.

He also published other articles in Italian, French and German besides Eng-

177 Dovring (1959), p. 1–11.

178 See Appendix 1. Bibliography, no. 42 with further references.

179 Dovring (1959), p. 1–2, pedagogically explained by Mellor (1966) p. 22–23. The difference in the rate of growth between the total labor force and the labor force in nonagricultural employment gave the growth of the rate of the share nonagricultural labor force. If the total labor force grew 2 % and the nonagricultural labor force grew with 5 %, the proportion of the nonagricultural would increase at the rate of 3 % – in numbers e.g. 200 in total labor force and 20 in nonagricultural, which is 10 %, would later be 204 of total labor force and 21 in nonagricultural, which would result in 10,29 %, which is a 3 % rate of growth of the share.

180 Johnston & Kilby (1975) p. 84–85, 123, and p. 226, 123 where later works of Dovring in connection with this issue is quoted. Hayami & Ruttan (1985) p. 19, 420 cite Johnston & Kilby with further reference to Dovring (1959). I am grateful to Ronny Pettersson for these quotations.

181 Dovring (1959), p. 2.

182 Dovring (1959), p. 9.

lish in the 1950s, as spin-offs from *Land and Labor*; his research was becoming quite well known in Europe. When he became a full professor in the U.S., he would further develop his earlier studies. For instance, in an extensive report for FAO from 1965, he discussed the problems of manpower in agriculture.

He also got the opportunity in 1965 to contribute to *The Cambridge Economic History of Europe* Volume 6 with a chapter about agriculture in the eighteenth to twentieth centuries.[183] According to Science citation index's measure of his impact until the 1980s, this was his most quoted shorter contribution. The chapter was not only a summary of *Land and Labor*, but was worked out as an extra investigation related to his earlier book. He added a new subject, technological change, studied and described as other subjects in the book with an immense quantity of examples.

In an introduction, he referred to and evolved his earlier investigations. He also tried to come down to the ultimate frames for European agriculture. For instance, he declares that with a low efficiency level the maximum cropland that a man can tackle is soon reached.[184] In England this was 10–12 hectares per man; in other countries only 5–6 hectares per man. These restrictions meant that a more abundant land supply usually meant more grazing and more animal production.[185] Another far-reaching consequence was that France and Germany in the sixteenth century had already devoted most useful land to cultivation; later clearances mostly referred to earlier cultivated land which had been abandoned or was being used more extensively.[186]

He had a wonderful little worldwide summary of how the dominating crop-regime drives the choice of implements. Most of Africa had manioc, yams and sorghum, requiring massive hoeing, but not the plough. Pre-Columbian America had maize and potato that could produce quantities of food from a small area, but mainly was cultivated with hand tools. Also the ancient civilizations in Asia had a wide array of cultivated plants that gave occasion for putting massive resources of human labor to intensive work. In the contrary Europe only had a restricted choice of plants, and these required plowing and harrowing, sowing by broadcast, harvesting with scythe and sickle. There was no real possibility of intensification. And most of the work had to be completed during a short season, leaving long off-seasons for stock-rearing and other activities.[187] This steered much of Europe's specific potential in historical development, but Dovring did not exploit his overview further. Instead he dove into details of technological change.

The main part of the article consists of a description of how new plants, implements and methods were introduced. He uses the same method as he had done

183 Dovring (1965b).

184 Dovring (1965b), p. 612.

185 Dovring (1965b), p. 612–613.

186 Dovring (1965b), p. 618–620.

187 Dovring (1965b), p. 632.

before, and presents tables of the number of machines, or the amount of fertilizers per hectare used in different countries in Europe at different periods. But he was not able to present the basic data in the same way as in *Land and Labor*, and thus his statistics in the chapter in *Cambridge Economic History of Europe* are, though interesting, not as transparent and useful as the tables he gives in his book.

Also evident are those strange omissions of key authors that he had made before. When Swedish research about agrarian change in the last decades is enumerated in the list of literature, one book is missing, namely Gustaf Utterström's imposing dissertation published 1957. This is one of the most exhaustive books about agrarian change prior industrialization. But Utterström had criticized Dovring some years earlier in a review of the textbook about agricultural history.

Turned into administrator

Dovring's research about the recent history of agriculture gave him a chance to secure a job even before he had published his major work. He had spent a lot of time in the library at FAO, and he was eventually offered a job in the organization. He rather soon realized that this was not a research position, as the mission of FAO was undergoing change with less emphasis on research.

A look at his publication list shows a sharp drop during the years he worked in Rome, and, for Dovring publishing seems to have been the essence of life. He did publish however, but contrary to his liking, not under his own name.

To this day, among his former colleagues at The University of Illinois, there is still a saying that Dovring used to assert that not even "wild horses" could drag him back to Rome and FAO, even though he was offered to return there occupying a leading position. The best source is however the letters to his mother, where he explained what he felt. We can follow his changed emotions against Italy, and his growing admiration for the U.S.

He had written to her previously in December of 1950, telling about his journey that year to Rome, Paris and London.[188] He penned with warm sympathy regarding the kind Italian people and their delicious food. This was before he was employed at FAO.

Between 1951 and 1956 there was a rupture in the contact between Folke and his mother. This was the period when she lived with the older brother, who Folke detested. Folke, thereafter, started to send postcards from places where his duties forced him to travel: the U.S., India, Yugoslavia, etc. Later he resumed longer letters.

At this time, he seems to have already focused on a future life: "USA is a country where one can live."[189] And, in a letter from December 1957, he declared that Americans were the most "grown up" among the people he had met, especially

188 PA-MS, Letter from Folke Dovring to Naemi Ossiannilsson, 17.12.1950.

189 PA-MS, Postcard from Folke Dovring to Naemi Ossiannilsson, 18.2. 1957.

compared with "childish" Italians.[190] The Americans were also "free from traditions".

In the same long letter written in December 1957, he told that he had to do a survey about agriculture and forestry in the Mediterranean area. "The idea was Myrdal's, and it was the last one that he suggested before he left ECE and went to India on his new assignment." Gunnar Myrdal had started to do research for his *Asian Drama*, published in 1968.

On his way to India, Myrdal had passed through Rome, and suggested the new project, which landed on Dovring's table. The result was that Dovring had to handle a group of dynamic and tiresome experts. Among those who visited him at this time were the later quite famous agricultural historian Ester Boserup and her husband. The Boserups had worked with Myrdal in Geneva, and they were now contracted to work with him in India.[191] Dovring had made their acquaintance years before, in Geneva, through Myrdal's mediation.

In December 1958, Folke again wrote a letter and complained about his work.[192] He had started a study on overpopulation, and did not want to be disturbed with administrative problems. But the weeks passed by, and he had to take care of experts and consultants from all over the world and organize their travels. He again talked wistfully about the U.S., where Karin, his wife, worked at Yale and commuted to Rome. Dovring tells his mother that "My winter (semester) there [in the U.S.] two years ago was perhaps the best time ever experienced".

Another year went by, and in August 1959 he wrote a long letter to his mother. It starts with a poetic picture of Rome:

> "The heat stands white and dazzling above the rusty brown mass of houses. The haze is steaming, dense and dusty over the parched soil of Champagnes. It has been like this for months now. There have been hardly a drop of rain, the tempests of Northern Italy has not reached Rome. The few sprinkles of rain cannot release the electricity in the air. The Sirocco blows more often and longer than usual, and it brings sand from the Sahara, which penetrates into the apartments. The Romans, superficial as always, are more uncontrolled and unpleasant than ever when the Sirocco blows."[193]

Dovring seldom revealed his capacity to write in a literary style; normally he expressed himself in a scientific and "boring" way. Evidently he was upset, and in this letter his distress sought an outlet. He could not sleep because of the heat,

190 PA-MS, Letter from Folke Dovring to Naemi Ossiannilsson, 21.12.1957.

191 Ester Boserup in her autobiography describes how she went to India in 1957 with Myrdal, but then in 1960 broke with him because of diverging opinions on development theory, see Boserup (1999), p. 18, 31.

192 PA-MS, Folke Dovring to Naemi Ossiannilsson, 15.12.1958.

193 PA-MS, Folke Dovring to Naemi Ossiannilsson, 15.8.1959.

and in the morning he had to: "check proof-sheets, tell half a dozen secretaries what to do and check that urgent correspondence is sent away before the plane to Baghdad leaves, and organize a comparison of important texts before they are sent away to five continents for comments." To manage this, he had to organize his day closely.

He did not mention a word about doing research. But Philip Raup, who was at FAO doing research in the library during the fall of 1959, said in an interview that he often sat together with Dovring in the library, and they talked quite a lot about research.[194] (Raup was a distinguished American professor in land economics; he was the one who later recommended Dovring to University of Illinois.) He had had the impression that Dovring was employed as a scholar, not as administrator.

In Dovring's December-letter he also told his mother that he had a hope – he was soon to be sent to America again, first on a shorter assignment, and then perhaps to stay at the New York office of FAO.

To the United States

In October 1959 Dovring again wrote his mother from the U.S.A.[195] He had spent some time in Washington and in New York, where he had met people who knew his book. He had been invited to give a lecture in Minneapolis at the University of Minnesota (certainly by Raup), and would also visit several other universities. Then he revealed that he had received "offers to be a professor at different universities", and told his mother "I cannot deny that I am tempted to stay here".

At Christmas 1959, he wrote his second letter from the U.S.[196] He had given lectures at the universities of Minnesota and Illinois, but the peak of the journey came when he was offered a full professorship at Illinois in agricultural economics. "The offer is so tempting that I have accepted."

He then speculated about his future: "Nobody forces me to stay at that University the rest of my life", and "one should not stay at one place the whole time". If he were to return to a U.N. career later, he would have more influence "as I will have been a university professor and will have written more books".

From a different view-point Dovring's future head of department, Harold Halcrow, has described how he came to hire Dovring.[197] He was at this time remolding the whole Department of Agricultural Economics at the University of Illinois and was looking for the best scholars he could find – a story I will come back to later. Halcrow wrote to Dovring (later he remembered how surprised he

194 Interview with Philip Raup, 27.6.2002.

195 PA-MS, Folke Dovring to Naemi Ossiannilsson, 24.10.1959.

196 PA-MS, Folke Dovring to Naemi Ossiannilsson, 23.12.1959.

197 Halcrow (1998), p. 174–176, and interview with Halcrow 24.5.2002.

was to get an immediate answer), not realizing that Dovring already was in the U.S.[198] Everything happened fast and quite informally. When Dovring gave a lecture in Urbana-Champaign in early 1960, he was already engaged as a full professor in land economics there.

Philip Raup had known of *Land and Labor* even before he met Folke Dovring in Rome, and was responsible both for inviting him to lecture at universities in the Midwest, and for his ultimate hiring. He had advised Halcrow to go for Dovring with *Land and Labor* as an obvious proof of Dovring's standing. Raup's work in Europe for many years afforded him a background of knowledge about European scholars and their research.

General history

Dovring still felt himself a historian, even though he avoided having contact with other historians. For instance, he did not attend the world conference in history in Rome in 1955, as he, according to what Karin remembered, did not want to meet his former Swedish colleagues.

However, during his years in Rome, he produced a textbook of general history. The publisher was the same as the one who had published his successful book *Land and Labor*. The next book, *History as a Social Science*, published in 1960 is about the nature of historical research. He discussed how history is distinguished from other subjects as it is about "the basic difficulty of knowing anything about something, which no longer exists".[199] He also stressed the "unity of science", the methods used in historical research must be explainable to other scientists.

He certainly came back to his credo; a large number of documents have to be studied before conclusions can be drawn. All evidences must be tested, and statistical methods used to assure probability. Only then is the historian prepared for the question of what he calls "use of intuition in historical research".[200] He held intuition to be a half-conscious summarizing of probabilities, which may be exercised by historians. An intuitive statement is, however, not a proof; it must be tested. Dovring then initiated an attack on a specific type of historian vividly described:

> "As matters now stand, it too often happens that someone is 'appointed' an intuitive genius, by admiring teachers or associates, because he has a striking way of expressing the same (maybe ill-founded) assumptions as his admirers."[201]

198 Halcrow (1998), p. 174–175, interview with Halcrow 24.5.2002.

199 Dovring (1960), p. 4.

200 Dovring (1960), p. 17.

201 Dovring (1960), p. 19.

It is unfortunate that such a person be given the authority to judge the result of others. It is obvious that Dovring had a specific person in mind, and that person could be no other than his old enemy Erik Lönnroth, who fits the description well.

Dovring went on to attack one of the main tenets in the source-critical school in Sweden. They had separated "vestiges" on the one hand from "narrations" on the other, where the first kind of source was considered as the more reliable. Dovring held this to be an oversimplification; a source's use as a narration or a vestige depends on the aspect one chooses for the research.[202] Similar critical standpoints had been launched by several scholars. For instance, Dovring's great favorite Marc Bloch mentions this in his book about the historian's craft from the early 1940s.[203]

Dovring also discussed the historian's responsibility. The over-shadowing task for an historian is "to keep our minds free of unrealistic ideas".[204] He also talked about the danger of the "Golden-Age" myths in many countries; nostalgic dreaming of past glories.[205] This was a statement based on experiences from the catastrophe of the war, when historical arguments had been misused by dictatorial regimes.

His way of building a historical synthesis would be "by uniting smaller elements of information into larger units of narration".[206] This can be seen as a description of the method he had used in his earlier books and articles, both when he worked with the Middle Ages, and when he worked with agriculture in the twentieth century.

At the end of this book returned to the point where he had started in his dissertation one and a half decades earlier: the unique and the regular. Social history must work with the regular. The unique event can only be described as such when the regular is fully grasped.[207]

Dovring's textbook was used and reviewed in several countries and was translated into Japanese. Even in his native country, it was occasionally mentioned.[208] The reviews were, however, not all positive. One American reviewer, C. Vann Woodward, in *American History Review* 1961 compared him with Isaiah Berlin, the philosopher of history a la mode at the time, but concluded that Dovring did not have "the challenge and verve" of Berlin. In England, W.H. Walsh in *English Historical Review* 1962 held that Dovring's arguments against comprehending the unique event were not convincing.

202 Dovring (1960), p. 29.
203 Bloch (1953).
204 Dovring (1960), p. 90.
205 Dovring (1960), p. 91.
206 Dovring (1960), p. 74.
207 Dovring (1960), p. 95–96.
208 For instance Torstendahl (1971), p. 206.

The influential historian Sylvia Thrupp, from Michigan, editor of *Comparative Studies in Society and History*, in a review-article, however, in 1964 referred to Dovring with appreciation, because he had stated: ”not much can be expected in the way of empirical generalization in history outside the study of mass phenomena.”[209]

Myrdal's last offer

Dovring's success in the international arena also furnished his last chance to come back to Sweden, which could have been the ultimate triumph for him. But he did not dare to take the chance. Again this offer came from his old benefactor, Gunnar Myrdal. In the early 1960s Gunnar Myrdal had returned to Sweden. He had left the country in 1951, after a rather unsuccessful period in the government, as the secretary of the department of trade. He had been made the scapegoat in a press-campaign against the Social-Democratic government, and the prime-minister (who besides really disliked the self-confident professor) decided to let him go. In the early 1950s, Myrdal worked with European questions; his years as the director of ECE were not the most happy or productive in his life.

He gradually turned his interest from the industrialized world to the developing countries in the mid 1950s, and spent the second half of that decade in India. After writing a major work on the South-Asian subcontinent (*Asian Drama* 1968), he was given the opportunity to build his own institute in Sweden at the University of Stockholm. The institute would be a research program dealing with international economics. The institute had no responsibility for the education of undergraduate or even graduate students. Myrdal had started to gather a competent group of collaborators, and then he remembered Folke Dovring.

Occasionally they had been in contact during the years. For instance in 1957, Myrdal in a letter thanked Dovring for valuable comments and sent Ester and Mogens Boserup, his collaborators, to visit Dovring in Rome.[210] When Folke obtained his position in Illinois, he immediately informed Gunnar about it. He told Myrdal that this position appealed to him, as FAO “has moved farther away from the conditions of a research institution than ever”.[211] The following year, Gunnar Myrdal wrote a letter of recommendation for Dovring where he declared that he considered him as “one of the most original minds I have come across”.[212]

During the summer of 1962, Karin and Folke Dovring made their last visit to Sweden. Folke had started to work on his third revised edition of *Land and Labor*,

209 Thrupp (1964), p. 100–101.

210 MA 3.2.2, Gunnar Myrdal to Folke Dovring, 15.4.1957.

211 MA 3.2.3, Folke Dovring to Gunnar Myrdal, 15.3.1960.

212 MA 3.2.3, Gunnar Myrdal to Centre for Advanced Study in the Behavioral Sciences, 25.3.1961.

and had traveled widely in Europe. In July, Gunnar and Folke met at the new Institute for International Economic Studies at Wenner-Gren Center in Stockholm. There an offer was made to Dovring to join as "visiting research professor", for a year, to study anything he liked, beginning in 1963.

Negotiations with Dovring were performed by another scholar Myrdal had brought home to Sweden from the U.S., Göran Ohlin. He had been an assistant professor at Stanford 1956–59, and an associate professor at Columbia. He and Folke had recently met at a conference in Bloomington, Indiana. (Ohlin was the nephew of Bertil Ohlin, a world famous economist, a friend of Myrdal and, at this time, the leader of the liberal party in the parliament – Sweden is a small country.)

Dovring answered in a positive tone in October; yes, he would like to come back to Sweden to the new institute, which seemed to be an "attractive environment".[213] A formal offer was immediately sent back in English to Folke Dovring. He answered that the only problem was that Karin Dovring did not want to come along for a whole year. Göran Ohlin regretfully replied that the institute could not pay her commuting.[214]

After a while, Myrdal started to be impatient and, in December 1962, he wrote to Dovring, asking if he could come for a meeting in New York in January.[215] Dovring answered in a letter in early January, still hesitating. He was tempted to work with "you and your associates again" and "touched by your wish to have me back in Sweden". With the letter he attached a proposal for research (about trade and food supply). At the same time he raised the level for what he would accept. A one year appointment was inadequate. Instead he wanted two years, and an "assurance of a permanent position in Sweden" after he had finished his two years.[216]

At their meeting in New York, Dovring declined Myrdal's offer. "If most of the academic world in Sweden were like you and your associates, I would have no hesitation at all." However his previous experience pointed in another direction.[217] He did not want to put himself "at the mercy of the academic system in Sweden". The only position he could accept was an offer to be a full professor permanently, and this was something that Gunnar Myrdal could not guarantee. Karin felt the same, and in contrast to other countries "there has been no perceptible reaction to her published works in Sweden". She would not go into negotiations with Stockholm University (which Myrdal had suggested) until there was some sign of "spontaneous interest" from someone at the university.

In March, Gunnar Myrdal made his last attempt; he and Folke had a long

213 MA 3.2.3, Folke Dovring to Göran Ohlin, 7.10.1962.

214 MA 3.2.3, Göran Ohlin to Folke Dovring, 15.11.1962.

215 MA 3.2.3, Gunnar Myrdal to Folke Dovring, 21.12.1962.

216 MA 3.2.3, Folke Dovring to Gunnar Myrdal, 2.1.1963.

217 MA 3.2.3, Folke Dovring to Gunnar Myrdal, 24.1.1963.

conversation over the telephone. Afterwards, Gunnar summarized that he was able to give Folke a two years' appointment as research professor. Indeed he could not give any formal guarantee for the future. However, personally, he had no doubt that Folke Dovring could either, if he liked, continue in the Institute or obtain a professorship somewhere else in Sweden.[218] "We are intent upon enlarging the higher personnel, at our universities very much." Then Myrdal made some comments that probably offended Dovring:

> "Generally speaking I have the feeling that you are thinking too much in terms of security. There is a tremendous scarcity of good economists in Sweden and in the whole world, and I see no real risk."

In addition, Gunnar Myrdal promised to do his best to support Karin Dovring. He was also "prepared to have any agreement on cooperation with the University of Illinois", and thought that he could have some influence by writing to them. The letter ended with a rather harsh comment: "Under these circumstances, it is up to you to make your decision. Nobody can do it for you. Write me a letter."

Dovring's answer was fairly rude: "I am slightly embarrassed because I thought that I had already replied to the same things which you bring up", and: "Maybe I did not make myself clear enough. The gist of what I said was that I could not accept your offer under present circumstances."[219] Dovring then explains that he did "not trust the Swedish University system, and I do not want to be at its mercy". And he also tells Myrdal that he has accepted another offer, to work for the OECD. (This work was however not substantial enough to take leave from University of Illinois.)

After this, all contact was broken between Dovring and Myrdal. They did not see each other, even though Myrdal visited the University of Illinois, and they never again exchanged letters.

In a way, Dovring let Myrdal pay for what others had done to him. Yet from Folke and Karin Dovring's standpoint, the idea of going home to Sweden more and more seemed a nightmare.

Dovring stayed in Urbana, Illinois and actually never even took a sabbatical semester. He explained to his wife that one could never be sure that the chair was not gone when you came back.

Yet a historian

Dovring's textbook of 1960 was not his farewell to history. In several articles and reviews written during his first decade in the U.S., he kept in contact with his old subject, while his main interest gradually drifted away to other subjects. In the

218 MA 3.2.3, Gunnar Myrdal to Folke Dovring, 5.3.1963.

219 MA 3.2.3, Folke Dovring to Gunnar Myrdal, 29.3.1963.

early 1960s, his historical research was mainly connected with his reediting of *Land and Labor*, and with his work on the chapter in *Cambridge Economic History of Europe*.

He also wrote reviews and surveys and published in journals of good repute such as *Agricultural History*, *Journal of Economic History* and *Comparative Studies in Society and History*. (He continued to write reviews in the Swedish *Historisk tidskrift* until 1956.) Apparently his contribution was esteemed, as he later was offered a seat on the board of *Journal of Economic History*, which was a very prestigious position.

Dovring became a member of the board in 1969 and was appreciated as such. In 1974 he was asked to sit for a second term as member of the board. In a preserved letter, the editor wrote in 1977: "as usual I will value your referee opinion of it", when he asked for Dovring's opinion about a received article. Dovring had thus acquired a real position of authority in the American Society of Historians by deeming over the publication of articles in this well-renowned journal.

In his review articles, he did not present as many new and thought-provoking ideas as he had in his more scientific work. He appeared as a firm adherent of Malthusianism,[220] but also declared that one of the main reasons for population growth in many countries was the peace and order, health control and developed economy that colonialism had established.[221] This was at the time conventional wisdom, which has however since somewhat eroded.

In the *Journal of Economic History* he presented the two important books that he felt belonged to the definite break-through for history of agriculture in the early 1960s: Slicher van Bath's book about European agriculture in a long-term perspective (*The Agrarian History of Western Europe*), and Ester Boserup's book about population and cultivation (*The Conditions of Agricultural Growth*). His estimation of these books shows again that, for him, the collection of a huge body of information was more important than a brilliant hypothesis built on a fragile base.

Slicher van Bath's book was very much favored by Dovring. In the review of 1965, he was impressed. That was not usual for him. His praise of the author is nearly overwhelming: the book is described as having scholarly depth, a grand design, to be the result of impressive research, and overall to be a very important contribution. Of course Dovring had several critical remarks. He was especially concerned about how sources can be better understood and utilized. Dovring hinted about his own earlier works in the field, and the review ended with a reminder of the "basic unfinished tasks" to which historians of agrarian society should address themselves.[222] It is as if a strain of longing for his abandoned field of long-term history comes over him.

220 Dovring (1961), p. 599; Dovring (1962), p. 370.

221 Dovring (1961), p. 606–609.

222 Dovring (1965b), p. 297.

Boserup's book was another thing. His 1966 review was negative. Boserup argued against Malthusianism. She contended that there had always been enough inventiveness to cope with a larger population. The Dovrings, Karin and Folke, had been on visiting terms with the Boserups when they belonged to the Myrdal-team, but Borserup's book was not at all to Dovring's liking. In his review, he declared that it was too speculative, had too few evidences and too much conjecture. In his opinion, Boserup was "fettered by a single idea".[223]

In 1969, Dovring published two articles with more theoretical aspirations. One of them, in the *Journal of Economic History*, was what he called a "non-dialectical theory of progress".[224] With this he meant a theory that did not postulate a necessary sequence of events. This has, however, nothing to do with dialectic reasoning in the common sense of that concept. Dovring most probably wanted to establish some distance from Marxism.

His main idea was that conventional productivity measures understate the rate of change. For instance, if we use a utility-function, one must ask how many clerks would be needed to do the job of computers; conventional economics cannot measure this. The real change in productivity is an accelerating progress, where new inventions follow each other in ever more rapid sequence. This theory of acceleration actually points to affluence as a problem as opposed to scarcity as a problem. This idea would steer much of his later thinking.

In a comment, Thomas Kuhn questioned Dovring's accelerators in this process of acceleration.[225] Dovring suggested that technological change was brought about by more generalized ideas, and a better grasp of reality. Kuhn instead argued that untutored craftsmen often had made the most important innovations in history. He also posed that science would prove only later that these methods, from a theoretical point of view, were the best. Dovring in his comment on the comment argued that the craftsmen used their intellect and that acceleration occurred "by 'tooling up of the mind' is a hypothetical framework".[226] The discussion probably had its root in the superficial way Dovring had presented his theory. As he did not elaborate on this theory of history it was seldom quoted.

The other article was a comment Dovring gave on an article by the leading agricultural historian Slicher van Bath in *Agricultural History* 1969.[227] Slicher van Bath was against the concept "agricultural revolution" as a label for the change of European agriculture; he assumed a more evolutionary and differentiated change. Only in the nineteenth century did most of Europe follow the British and American mechanization of agriculture. Dovring, in his comment, declared that he was of the same opinion as Slicher van Bath. He even sharpened the arguments. In-

223 Dovring (1966), p. 381.

224 Dovring (1969), p. 422.

225 Kuhn (1969).

226 Dovring (1969b), p. 432.

227 Dovring (1969c).

novations in agriculture were made one by one, with long intervals of no technical change. Europe never went through an "agricultural revolution" in the eighteenth and nineteenth centuries, and England's role therein is typically overestimated.

This comment by Dovring got some attention, and has been quoted even quite recently, in important textbooks. For instance, Daniel Vasey in his *An Ecological History of Agriculture* of 1992, discusses the concept "revolution", referring mainly to Dovring. Vasey declared that he sympathizes with Dovring's objection to the word "revolution" because change was so slow, but Vasey declares that he still uses the word because clear alternatives are lacking.[228] Another example can be taken from Immanuel Wallerstein, who, in his famous synthesis about the world system, referred to Dovring's skepticism in regards to "revolutionary" change in the agriculture of Western Europe and especially of England. This underscored Wallerstein's explanation of the British success having been a result of geopolitical domination, rather than internal development.[229]

Now and then, throughout the 1970s, he continued to write reviews of books on history, but rather seldom. His mind was occupied with other subjects.

Dovring's impact as historian

In Sweden Dovring was one of those who came to forge a change in medieval historical research, and his books are still much used as highly informative and reliable. His influence was however disguised, as his attack on Erik Lönnroth had made him a non-person in the society of Swedish historians. Evidence of this is the fact that his *kasuistik*-method was used, but it was never openly declared as founded by him. In the long run, this dominance of a group of persons around Lönnroth came to impede research about the Middle Ages, as they tried to deflect questions or hypotheses they disliked. Dovring was only one, if perhaps the most flagrant, example of persons who were pushed aside as they did not acquiesce to the leaders. This does not contradict the fact that Lönnroth genuinely tried to raise funding for research in the Middle Ages, though, for different reasons, medievalism in history experienced an ebb in interest from the 1970s and well into the 1990s. (In other university subjects, as in archeology, the interest for the period increased during these decades.)

Internationally, Dovring's works met another fate. His work about agricultural change, from the late nineteenth to the middle of the twentieth century, came to be a reference book for international research. It was used much as his medieval books were used in Sweden, but without the connotations that his per-

228 Vasey (1992), p. viii.

229 Wallerstein (1989), p. 18. Also economists agreed in questioning the use of the word "revolution", e.g. Hayami & Ruttan (1985), p. 61 on the "green revolution" with reference to Dovring's discussion.

sona had in Sweden. His impact was, of course, far less strong than international top-scholars, but he had a larger reputation and his writings were more influential than any fellow Swedish historians of his own generation. In the 1960s, he was actually one of the more successful Swedish scholars in the international arena (though he then considered himself as an American). Even into the early 1980s, Dovring's *Land and Labor* was quoted extensively.

When he left history and, especially in the 1970s, took up more politically burning questions, such as the oil crisis and environmental care, he tended to be more isolated, and his voice did no longer reach out. In a later chapter I will try to answer why his work as an economist got a different response than his historical research. (See also Appendix 4 by Gabriel Söderberg.)

First I will continue with a main theme of this book. If we understand his relocation, physically and intellectually, from Europe to the U.S.A. as a process of push and pull, the first part of the book is about why he was pushed out from Europe. Next part is about how he was pulled over to America.

CHAPTER 5

Involvement in the expansion of higher education (1960–1987)

Why the United States?

Why did Dovring choose to go to the U.S.? His French was at least as good as his English. He had mainly worked with European issues, and he was rather well-established in the European community of scholars in the field. And why did American universities try to hire him when they had many good scholars nearby?

This chapter provides a context for the academic situation around 1960 when Dovring went to the University of Illinois. The text starts at the national and international level, and goes from present to past. It closes at the level of the specific department, down to the single teacher and researcher.

Dovring was not the only one to leave Europe; a wave of intellectual immigrants exited after World War II, as the United States at this time had established some superiority in higher education. When the French scholar J.J. Servan-Schreiber wrote *The American Challenge*, which was a best-seller and caused a far-reaching debate in Europe in the late 1960s, he pointed at the system for higher education as being one of the major causes which gave the United States an upper hand in industrial and economic development.[230]

Europe tried to catch up. Several countries in the 1960s and later introduced elements from the American university system. However, universities in the U.S. have continued to lead. This lead has lasted for one half to three quarters of a century. During the later part of this period, the international academic society has formed as one unit, with the U.S. at the top of the hierarchal pyramid.

The U.S. dominance today

The former dean of liberal arts at Harvard, Henry Rosovsky, 1990 put it: "fully two thirds to three quarters of the best universities in the world are located in the United States."[231] His estimation is in accordance with recent ratings of the top 50

230 Servan-Schreiber (1968), p. 67–73.

231 Rosovsky (1990), p. 29, cf. Graham & Diamond (1997), p. 10 on how he estimated this proportion.

and top 100 research universities in the world.[232] Certainly quite a lot of the worst colleges in the world also are located in the U.S. and, discussing the American university system, one must never forget the enormous diversity in the system. Here I will leave those thousands of colleges aside, and only discuss the one to two hundred research universities, which attract foreign scholars. Of these at least fifty can be counted among the top-universities in the world.[233] The University of Illinois, to which Dovring went, was and is ranked among these.[234]

The United States has the most foreign students in the world. Other countries with many foreign students are Britain, Australia and Canada. Even though France and Germany also have many foreign students, in this respect, they are losing position in the area of higher education.[235] A main reason for students going to the U.S. and other English-speaking nations is that these countries have an advantage above other countries because English is the lingua franca all over the world. After the 1950s, all other languages with such ambitions have been giving way: German of course, but also, though under resistance, French, and other major languages such as Russian or Spanish. In a global society one language has to dominate, to facilitate communications.

The reasons why English became the dominating language has historical roots in the military and economic supremacy of, first, the British Empire in the nineteenth and early twentieth century and, then, the U.S. However, to concentrate only on military and economic power would be to simplify.

The first question is whether the United States today really holds the leading position because of military capacity. Most examples from later history tend to show that a strong nation in a military conflict can win the first rounds of the match with sheer power, but the whole match will be lost without support from the common man. An ongoing democratization of the world; an increasing role of the broader strata, has been a main feature in world history at least the last two hundred years.

The other question concerns economy. The economic hegemony of the U.S. was a fact directly after World War II, but today it tends to diminish relatively when countries in the rest of the world – not least in Asia – rapidly increase their production.

A field where the influence of the U.S. is still overwhelmingly dominant is popular culture. American movies and TV-programs are spread virtually everywhere, and they give a certain, American, view of the world. The music is formatted in an Anglo Saxon style, which not only concerns the words and the texts but

232 In an often-quoted ranking from 2003 over 70 % of the 50-top universities in the world are situated in the U.S., and of the 100-top universities over 60 %, see Appendix 2.

233 Rosovsky (1990), p. 36.

234 In the early 1960s it was among the top-ten, but today it is ranked around 20–40, see Appendix 2.

235 Barber (1992), p. 1020.

also ways of expressing a mentality (at a glance mostly about love, but actually also about other feelings and values). Computer-games is another example. To this a whole life-style is connected, with food, drink, clothes, etc. These cultural and social elements taken together carry a tremendous impact on the mentality of the world population.

Just as importantly, the influence on the emerging world-mentality is also determined by the intellectual elite, particularly in science and in university-culture as a whole. The need for a common language has certainly paved the way for English here, to a larger extent than in nearly any other branch of society. The tendency to formulate ideas and theories in English is strengthened by the fact that the world-wide community of academics is turning its face towards the centre, which is the United States.

However, the American superiority in higher education is caused mainly by factors other than English as the common language of the world. When I discuss these factors, I will start with the contemporary situation and, thereafter, take up historical causes. The world leadership position of American universities is rather recent, experiencing a crucial period around 1950–1975. Research about universities in the U.S. is an enormous field and I do not intend to take this overview up to date, but concentrate on the period from the 1950s to 1980s.[236]

I will later use the Department of Agricultural Economics at the University of Illinois, and Folke Dovring, as an example, which serves to mirror and explain the larger process of change. This also gives an explanation as to why Dovring and many other scholars from Europe chose to go to the U.S. and why they were needed in that country.

Causes for American dominance

American and European universities have the same heritage; the differences shall not be overestimated. Their common university system is today the standard for universities world wide. Admittedly, a more wide-ranging discussion ought to include the rest of the world. Nonetheless, I am going to concentrate on the reasons why America is in the lead compared to Europe, and my list is far from exhaustive.[237] Top universities in the United States have during most of the period after World War II:

1. Been more diversified, and have less interference from politicians.
2. Given more power to the president of the university.

236 In my short overview I mainly follow: Kerr (1982); Bok (1982); Bok (1986); Rosovsky (1990), Geiger (1993).

237 Such lists are not uncommon, and for instance Trow (1991) points at: widespread regard of higher education; little influence from government: strong presidents; continuing education; flat academic hierarchy, university service to society, etc.

3. Been more market oriented / more competitive.
4. Had scholars who are more transferable: the nation is one pool of scholarly resources.
5. Had more students: a larger proportion of the population attends higher education.
6. Had more money, both from public and private sources.
7. Been to a certain degree marked by the "Mark-Twainian" freedom.
8. Been at the top because they are at the top.

1. *Diversity and political influence.* There are few national standards, whereas in Europe every nation has developed detailed principles for universities and colleges. This encourages flexibility and diversity in the American system.[238] Rosovsky argues that the freedom from state-interference leads to decisions less determined by other goals than those promoting the university as such.[239] On the other hand, heredity in higher education is a strong force. European universities often have hundreds of years of traditions to lean against, which serve as a barrier to political interference. In the U.S. the central government has an influence through grants. The differences should thus not be overrated.

A most striking effect of the flexibility in America is the great number of low-standard colleges, many of which would never be allowed to exist in Europe (as certified education). Bok argues that these low-standard colleges have an assignment – to make some sort of higher education available for people who otherwise never would have gone further, or even to fill up the gaps from their secondary school.[240]

An important part of Anglo-Saxon university-culture, both in England and in America, is the large number of private institutions. Considering the role of tradition in the university system, this role of private universities will probably never be adopted by most European nations.

2. *President and faculty.* Normally in the U.S., the board appoints the president of the university, and the president appoints deans and most of the administration. In Europe, the faculty most often elects the "rector" and the deans.[241] (This is changing today.) However, when it comes to the actual influence over the governing of the university, the difference is less pronounced. The faculty in American universities

238 Bok (1986), p. 10–11. It must however be mentioned that diversity demands strict but restricted conformity, to make the system function on a national scale, with for instance SAT-tests or the like for all students, or transferable "credits" for graduate students.

239 Rosovsky (1990), p. 33.

240 Bok (1986), p. 31–32.

241 Rosovsky (1990), p. 33 emphasizes this difference, but also is quite clear about the fact that the president relies on the faculty cooperation, p. 205. Also see Kerr (1982), p. 26 about the influence different groups wield in universities all over the world.

today has an influence on a number of questions (as hiring of new members of the faculty). Indeed, it is nearly impossible to direct a department or a university without sustainable support from below. The days of the more autocratic university president in the U.S. mainly belong to earlier periods,[242] before the country became the leader of the academic world society. Generally, the rule is that the more successful the faculty is, in science and education, the more influence they will have. The power of American university presidents and chancellors as a factor for success has probably been somewhat overestimated in the discussion.

However, the top administrators of American universities still tend to have larger degrees of freedom than their colleagues in Europe in, for instance, appointing star scholars and giving them top wages.

3. *Market and competition*. Every university acts on some kind of "market", in the sense that students and teachers have a free choice of where to go (if it is not a very small discipline in a small country). American universities however, tend to act, and conceptualize their actions, more as if they worked in a corporate arena with students and fund-givers, which makes them more sensitive to demands.[243]

Important to note is, however, that most universities do not have high profit as a goal, they rather strive for recognition and prestige. In order to get prestige they have to deliver excellence, competing with other leading universities. They also have to deliver what students, alumni, corporations, etc. want. This is necessary to obtain donations, private funding and student fees. This is a delicate balance. To rely only on peer-rating of academic performance would not lead to successful fundraising from the public. To adapt only to commercial interest could undermine academic excellence.[244]

Derek Bok has pointed to the relative freedom from government control, and thus the absence of a national standard, as one of the main causes behind the sharper competition among American universities. A fixed hierarchy among universities, as often is the case in a state-organized system, is replaced by a number of ranking

242 Kerr (1982), p. 33–34.

243 See Trow (1989) about this difference. Bok (1986), p. 18 uses "marketplace" as a figure of speech. In a discussion about commercialization of American universities Bok (2003) argues that the reason mainly is the growing importance of science in a knowledge-based economy, but he also points at problems and perils. This topic is, however, not within the province of this book on Dovring and the university system up to the 1980s.

244 See Trow (1989), p. 390 about the British universities, without friends they were mistreated (during the Thatcher-era), where universities in the U.S. has many friends and supporters in the society. Geiger (1991) p. 204 about the academic oligarchy in the prewar U.S. with shared values, but "it proved difficult to expand support for research". The debate about why and how the American university system is endangered by commercialization has been rather intense the last few years. See Bok (2003) in footnote 243, and for a more radical anti-commercial position, see Donoghue (2008).

lists. Universities in the U.S. are, from a European standpoint, obsessed with ranking, but that is a necessity in a competitive system such as the American.[245]

The importance of ranking has dramatically increased world-wide in the last five to ten years, which is a clear sign of a formative phase in the shaping of a global community of universities. This will give the U.S.A. a leading position, and, at the same time, challenge this position. Other nations will try to catch up. (However, this world-wide ranking is still rather crude, and as it is currently constructed a threat to humanities.)

4. *Scholars mobility.* An element, which encourages competition, is the enormous academic pool that the United States of America forms as a nation. This advantage may only be realized if many individuals move freely throughout the nation. That is the case. Most scholars will pass through several universities during their career, and at the top research universities in the U.S. a new doctorate has to leave the university where he or she graduated to continue his/her career somewhere else.

This is partly due to the American way of life, at least among the elite, where many people move several times during their life. This mobility can be seen as a way to keep this enormous landmass together as a nation, and it influences many institutions, not only the universities.[246] However, the faculty mobility is a quite recent phenomenon. Before the 1950s, many major universities recruited their staff from their own graduate students. This shaped a system characterized by local pride where outsiders could be seen as a threat against the institution. The breaking up of the inbreeding continued for a long time, and was one of the prerequisites for the establishment of a national market for scholars. Once established, this selected and sorted the scholars, so the best went to the best universities. I will come back to the cessation of, and also the stigmatizing of, inbreeding around and after 1960.

In a smaller nation, or in a localized system, it is possible to know most of the important scholars in the field personally. When a huge pool of persons is formed, it becomes impossible to know all individuals and their qualities. Other ways to evaluate have to be developed. Articles in journals become partly advertisements and people have to display themselves before public at congresses. The peer-review system serves as the hub for the publishing of scientific articles.[247] These instruments for handling a large pool of scholars are not only sorting the American

245 Bok (1986), p. 15.

246 See Jencks & Riesman (1969), p. 191, about the connection between movable Americans and movable academics.

247 For a thorough critique of peer review see Chubin & Hackett (1990). Shatz (2004) reports all the investigations that have proofed the deficits, but at the same time, p. 12, alludes to Churchill and declares that peer review is the worst form of evaluation, except for all the others.

scholars, they are also becoming a world-wide system with the U.S. universities setting the standard.

5. *More students.* With few restrictions in establishing universities that have freedom to build new programs, there is no point in a restricted admittance on a national level. Admittance to the premier universities is, of course, restricted, but generally every one who wants to (and can pay for) can find a college to attend. Historically, the United States has an age-old respect and admiration for knowledge, where mass-based education has been as much the goal as selecting an elite. In Europe the main goal was to educate, and to sort out, an elite. Higher education for large parts of the population has only lately been a main goal in most European nations.[248]

This gave the U.S. for many decades a larger proportion of the population who continue to "the tertiary level" of education – to colleges and universities. The U.S. was, together with Canada, outstanding regarding the proportion of current college students relative to the population as a whole and in the number of students completing a college education.[249] European nations are catching up, and today this difference is no longer as important. It must also be admitted that the comparison is a little unfair, as the college-level in U.S. partly covers the Gymnasium – or lycée-level in Europe. The American undergraduate education is also often more general than the European.

Of certain interest is the life-style aspect of American mass education, with the importance of athletics, fraternities/sororities, etc. Certainly, even European students live out their first period of freedom as young students, but fewer of these experiences are furnished inside the university structure. By giving the undergraduate students (and their parents) a "package", the university can transfer money from students' tuition to basic research, thereby resulting in a higher peer-estimation.[250] This is a symbiosis between university research and mass education. The result is both a higher level of education and of research.

One further consequence must be considered. When a large proportion of the population has a college education, this becomes a prerequisite getting any job

---

248 In 1968 this was one of Servan-Schreiber's main arguments, and he argued that the heavy investment in education made in the U.S. was "a primary factor in economic development", Servan-Schreiber (1968), p. 71, cf. statistics p. 50.

249 Nearly all authors in the field mention that a larger proportion of the population attends the college-level, and the proportion can be counted in different ways. For instance according to Kurian (1997) (mostly based on UN-statistics), p. 295–298, 6 % of the population were students, more than half of those were 20–24 years old, and 40 % had studied at this "tertiary" level. In these three measurements the United States rated, together with Canada, in the top percentages, much above other countries.

250 Geiger (1991), p. 212 states that research funding during the problematic 1980s came from "increased tuition".

above the lowest levels. Educated people will occupy the leading positions, and they will, in turn, question why someone has not finished his or her studies.[251] This will result in influencing a large part of the population to attend higher education.

Another factor, which is peculiar to the American system, is the role alumni have in the system. For a large part of the population, their time at their college/university was one of the highlights of their life and it opened a career for them. They tend to revere and support their own university.

6. *More money.* A larger proportion of the gross national product (G.N.P.) is spent on higher education in the U.S. than in other countries; it has been so for a long time. (The same is not the case for education at the lower levels, where the U.S. spends less than many other countries, which has effects I here leave aside.) And, as the United States has a higher per capita G.N.P. than most other nations, this proportion will have quite an effect on the total amount of funding available. Especially since World War II an increasing amount of funds has come from federal and state sources. In addition tax reduction benefits afforded for grants and donations encourages additional, substantial contributions.[252] The effect of the "sputnik-chock" in 1957 was also immense, and resulted in a massive investment in research and development.[253]

The responsiveness of research universities has lately effected an expansion of academic cooperation with firms, companies and the federal government. Even if federal and state funding has slowed since the 1980s (and the tax-revolt), this does not change the general pattern for the last fifty years. An important source of income are also parents who are willing to pay quite high fees for the education of their children, of which at least some portion goes back into research.

In addition, one factor not to be forgotten, is that the middle class tends to vote to a larger degree than the lower classes in the U.S., and politicians are thus more sensitive to the demands from the middle class (than in most European countries). Among the issues favored by this stratum of the society, support to higher education tends to be on the top of the list.[254]

More money means better resources, for laboratories, libraries, salaries, etc.

Interlocking of factors 1–6. Many of these factors work together. For instance, more students will give more money, and esteem of higher education by the whole of society is an umbrella under which many of the above-mentioned factors could be included. These six causes are the ones most often discussed. The following two are not, probably because they are taken for granted.

---

251 Blecher (2002).

252 Rosovsky (1990), p. 31.

253 Geiger (1993), p. 161–166 on the challenge the Soviet satellite caused and p. 166 on the steep increase in research of G.N.P. for research around 1960.

254 Interview with Derek Bok, *Dagens forskning* 2003.

7. *Freedom of thought.* The whole Western world, and to a growing extent the rest of the world, experiences the right to think and publish freely. This is a fundamental prerequisite for all scientific and academic activity. Without the right to question everything, and freely investigate everything, the whole intellectual endeavor would stagnate and, in the end, fail. Academic freedom is closely related to the greater concept of freedom of thought, which, once established as a right for common people, is very hard to root out. Once the sweet taste of freedom has been felt, people will be ready to fight for it.[255]

America had, compared with Europe in the late nineteenth and early twentieth century, established a non-traditional life-style, which influenced their perspective of the world. I do not intend to suggest that American scholars, or the American people, had eliminated prejudices and idiosyncrasies. Rather, they had established other sets of them. Indeed, for a long time, the frames for thoughts in many institutions were as restrictive as in the old home countries.

Nevertheless a certain American non-traditionalist view was developed. I will call this the "Mark-Twainian" freedom. Mark Twain is considered the first American author of standard who did not care about Europe.[256] For him the Mississippi river was the centre of the world, and Europe was a strange periphery. Turning its back to a certain number of European traditions characterizes this specific freedom of thought. Certainly this Mark-Twainian freedom did not affect the whole of the American university system, but its influence was so large that a healthy disrespect for the Old World system of ideas did spread.[257] For instance, the historians' obsession with the "state" in Europe was not such a strong issue in America. The historians in the U.S. had other obsessions, such as the "frontier".

10. *The top stays at the top.* This last point is perhaps the most important of all the causes I have mentioned hereto. Even if it seems to be a matter of course, this factor had an enormous importance during a period when a world society of scholars was being formed.

The best want to be where the best are. If there is a lot of movement of faculty within the country, and if universities are competing to get the best scholars, you will get a situation similar to the one you have in professional athletics. The best football players want to be on the best teams (besides getting the highest wages there). A team or a university, which has established itself on the top, will continue to be there just because it is there. The position perpetuates the position. A fact is that, whenever a national league is firmly established, in any sport, many of the teams who belonged to the top league in the beginning will still be at that top

255 Quotation from Wittfogel (1980), p. 449.

256 Mark Twain (Samuel Clemens) later in his life, certainly went to Europe many times, and was inspired by its great tradition, but that does not contradict the interpretation of his approach, and my way of using it.

257 Similar arguments in Kerr (1982), p. 112.

league many decades later. Transferred to universities, this would imply that the top universities stay at the top just because they once had been established there. Harvard and Yale are American examples; in Europe there are the medieval universities of Oxford/Cambridge and different scholarly institutions in Paris, stemming from the Sorbonne. Many have noticed this peculiarity. Kerr has stated it as: "The scholar dislikes intellectual isolation and good scholars tend to swarm together."[258]

In a globalized world, the academic society increasingly works as one community. Universities will be compared worldwide. This turns out to be a comparison not only between universities but also between states. A pyramid is formed, and academics from all over the world prefer to join the leading universities in the leading nation. The United States established a position as the dominant nation in the academic society at the same time as this society became really global. This gave America the advantage of a huge number of scholars coming from all over the world to attend and develop the universities at the top of the pyramid.

This intellectual clustering in the centre of an empire has always been typical, as in China and the British Empire. The difference today is that traveling and other forms of communications have formed an academic and intellectual world society more interconnected than ever before. Today, with the internet-revolution, this fact is obvious to everyone. In such a society comparisons will foster competition.

Still the reason why the United States assumed a dominant position also has roots further back in history.

Historical causes before World War II

From the beginning higher education in America was more open to access. At the time of The Revolution, America had nine colleges, where England had two, Oxford and Cambridge.[259] Certainly the American colleges were of a lower standard, but it was important that they had been established. By the early nineteenth century, a college that had fulfilled the minimum requirements had the freedom to govern its own institution. This was the base of the unique American system.

An expansion came after the civil war, when land grant universities were founded in every state. The government gave away large tracts of land to public universities, and especially in the Midwest, the frontier at that time, these institutions became of major importance. They were meant to serve the practical life, and were strong in agricultural sciences and engineering, but at the same time they were to teach more scholarly subjects. The farmers who paid wanted their sons to get practical knowledge, but also to be able to hold their own against

258 Kerr (1982), p. 92.

259 Trow (1991), p. 376.

lawyers and other important persons.[260] Also daughters soon came to the university. Women were admitted to the University of Illinois as early as 1871, four years after its founding. The model these universities were built upon was mainly the German, where education and research were combined. This formed a new combination, giving respectability to research about daily life and practical matters.[261] Experimental farms and extension service were a part of this change.

From the 1870s and 1880s, the American universities developed modern research and the first American Ph.D.s. The quantity of original research soon started to be comparable with universities in the Old World.[262]

This advance of research was followed by a battle for academic freedom from around 1900 until World War I. The professors lost many individual battles; they were fired or subjugated. In the long run, however the faculty achieved their goal: to get academic freedom (to decide over curriculum and research) and tenure (lifetime employment) for professors. The faculty later gradually asserted more influence, and, generally, faculty gained strength as faculty gained distinction.[263] The march to the top of the pyramid was also a march towards academic freedom.

At the same time, America upheld superiority in the number of students attending college. For instance, in 1910 more than 300 000 students went to higher level of education in the U.S. compared with 40 000 in France, which at that time had a population comparable to that of the U.S.[264] In the 1920s, a further increase in the number of students followed.[265]

A factor of highest importance for American universities was the influx of intellectual immigrants, Jews and others, during the 1930s.[266] Before that, the United States had not acquired a leading position in world research. When dictators started to purge their own countries and surrounding countries of leading intellectuals, the balance started to turn in favor of the New World. I am not going into the fascinating story about this forced emigration, and the reasons that purged intellectual immigrants mainly went to the U.S., but it gave the American universities immediate contact with the cutting edge of research and scholarship.

260 Jencks & Riesman (1969), p. 4.

261 Kerr (1982), p. 47.

262 During the nineteenth century, before the 1880s, most Americans who wanted to do a Ph.D. that did not go to Britain went to Germany, see for instance Berelson (1960), p. 11.

263 Bok (1982), p. 4–5; Kerr (1982), p. 34; Jencks & Riesman (1969), p. 15. This issue is one of the most important in the history of American universities and has inspired a huge literature.

264 Bok (1986), p. 11; Trow (1991), p. 158.

265 Jencks & Riesman (1969), p. 77, statistics from this early period are somewhat contradictory, but two leaps forward are recognizable, around 1900 and in the 1920s.

266 For instance Rosovsky (1990), p. 31. The literature about this purge of intellectuals in Europe is enormous, not the least in Germany.

The crucial period 1945–1970

After World War II, the government decided to offer every soldier four years of paid education, with all basic costs covered, for fees, livelihood, etc. (the G.I.-bill). An unexpectedly large number of young men took part in the program. This gave the universities not only a larger number of students, but also an increased proportion of farmers' and workers' sons in the higher education.[267] This flood of new academics, from a larger intellectual pool of resources among broader layers of the population, came to determine further change. Many of them became teachers at the universities and when the leap for higher education came they were at hand.

The leap forward in the number of undergraduate students from the 1950s to the 1970s is of historical dimensions. Basic ideological and economic causes explain the change, which I will not go further into here.[268] In a retrospective view, this leap can be seen as a decisive change that prepared a basis for the society of our own time. In the U.S., the increase in the number of students came much before the European change in the same direction. Half the population went on to higher education some years after the war: the leap forward can be demonstrated in numbers, se table 1.[269] (As this is a biography of Dovring, working in the American system from the 1960s to the 1980s, I have deliberately left the last decades out of the table.)

Table 1. The leap in higher education in the U.S.

Year	*Students in higher education*	*Ph.D. doctorates*	*Faculty*
1947	2 300 000	2 900	
1950	2 300 000	6 600	247 000
1955	2 700 000	8 800	
1960	3 800 000	9 800	381 000
1965	5 900 000	16 500	
1970	8 600 000	32 100	573 000
1975	11 200 000	34 100	
1980	12 100 000	32 100	846 000

Source: Adkins (1975), Trow (1989); National center for education statistics.
Comment: "Faculty" counts different individuals, not reduced to full-time equivalent.

267 Kerr (1982), p. 52, the bill "sent a seismic shock" through the academic society.

268 Two main explanations are: the individuals demand for upward social mobility; demand for competence from the society, see Clark Kerr's introduction to Adkins (1975).

269 Based on Trow (1989), p. 371 for students and Adkins (1975), p. 193–195 for doctorates.

From 1955 and the following years, the number of students thus grew by about 50 % every fifth year. During the twenty-year period from 1955–1975, the total number of students grew by fourfold. (That these years also saw the student-revolution is a part of the story.) This increase shifted the balance between public and private institutions. In 1950 half of the students went to private institutions, but in 1970 only a quarter. The decrease in the portion of students attending private institutions flattened out in the 1980s to a level slightly over 20 %.

The doctorates have a slightly different curve. After a natural decrease during the war, from a pre-war level around 3 000, a sharp increase followed between 1947 and 1950. This was certainly as a result of the G.I.-bill (which did not have the same obvious effect on the total number of students). The increase of doctorates continued until about 1954–1955 when it flattened to a level of between 8 000 and 9 000. In the 1960s a new increase started. Between the pre-war level and 1970 the United States had a tenfold increase of the number of doctorates.

There is, of course, a correlation between the increasing number of students and of doctorates. On one hand, an increase in the number of students explains that the increase in doctorates starts five years later. On the other hand, more students needed more teachers. As I already have pointed out, many of them were at hand because of the G.I.-bill.

Faculty did not increase at the same rate as undergraduate students and graduate students in the 1960s.[270] Roughly, the number of undergraduates per faculty rose from ten to fifteen, and the number of Ph.D.s in relation to faculty doubled. In the 1970s, the Ph.D.-production per faculty decreased. Certainly, the 1960s was the most dramatic period when new faculty was much in need all over the country.

Parallel with this change, the funding of the universities increased. Around 1930, universities did not have many federal research grants and little income from philanthropic foundations.[271] During and after the World War II, this changed totally. From around 1940 to 1960, higher education got a rapid increase in funding from the government.[272] The growing amount of research money accelerated further in the 1960s. This was partly a result of the "sputnik-shock" in the late 1950s, and the following competition with the Soviet Union. The share of G.N.P. devoted to national basic research more than trippled from 1953–1968, rising from

270 The real number of the faculty is always somewhat floating. The numbers in table 1 are from Clark (1987), p. 12. In Stadtman (1991), p. 786 the following numbers are given: 1969–1970: 474 000; 1970–1975: 628 000; 1975–1980: 686 000; 1980–1985: 724 000.

271 Bok (1982), p. 24. Agricultural research however got funding from the federal governments much earlier.

272 Kerr (1982).

$440 million to $3 300 million.[273] After 1968, a decade of stagnation followed. Thereafter, higher education has seen a slow increase of research funding.[274]

The American system seems to have kept its fundamental structure through the great leap, which proves the flexibility in the system.[275] Indeed the magnitude of the increases did cause some important changes. The two most important were: the breaking of localism and inbreeding; and a brain drain from the rest of the world. With these changes the new system that elevated America to a leading position was established. It did have its roots far back, but did not take its definite form until after World War II.

Christopher Jencks and David Riesman devote a chapter in their study, from the late 1960s about universities, to the breaking of localism.[276] According to them, this was a process which had been going on for over half a century, especially during the last generation. It affected the graduate level much more than the undergraduate. To estimate the number of graduate students who finished their masters or Ph.D. in another part of the U.S. than where they grew up is difficult. The authors' estimate is that probably half of the graduate students by 1962 came from other states.[277] Another measure is the number that obtained their baccalaureates and doctorates at the same institution as where they had been undergraduates. In the late 1950s, as many as 60 % of the graduate faculty at the top universities had obtained their baccalaureates and doctorates from the same institution, and at the bottom of the ranking this proportion was only 15 %.[278] Thus inbreeding at this time was typical in top institutions. If we instead turn to the faculty, the top universities provided other universities with staff members, but also kept rather many of their own products. Across the whole system the proportion of inbred faculty was as high as 15 % in the late 1950s.[279]

During the 1960s, there was a definite in the inbreeding. The decade was characterized by an intensive competition for the best faculty. One faculty member in five changed campuses every year, and some "prestigious campuses routinely

273 Geiger (1991), p. 208–209; Geiger (1993), p. 161–166 on the challenge the Soviet satellite in 1957 caused and on the steep increase in the research of G.N.P. for research around 1960. Regarding the sputnik-effect see the immediate reaction reported in Berelson (1960), p. 2, 38. This is confirmed by many of the retired professors today.

274 Graham & Diamond (1997).

275 Trow (1991), p. 394.

276 Jencks & Riesman (1969), p. 161–189.

277 Jencks & Riesman (1969), p. 161–162, the main problem is that students fail to report where they grew up and instead register the town where they live. In 1962 6 % of the graduate students came from foreign countries.

278 Berelson (1960), p. 113.

279 Berelson (1960), p. 113–116. The proportion of in-breeding is somewhat too high, as many spent some years at other institutions before they returned to their home institution.

make close to half their faculty appointments by hiring them away from rival campuses".[280] A national market for graduates and teachers had been created. (How this was effected on micro-level, by deans, heads and faculty, is one of the findings in this book.)

This growing market for top scholars also influenced Europe. The brain drain was a hot topic in Europe in the 1960s. Servan-Schreiber has already been mentioned; another example is a book by D.N. Chorafas. A theme throughout the latter book is that scientists do not leave because they would command higher salary, but because they wanted to be where work that most interests them is being done.[281] The question was discussed in Scandinavia, and he quoted an investigation that said: "Sweden has definitely lost to America some top scientists", and this includes doctors, mathematicians and engineers.[282] The examples I have mentioned, Dovring and Borgström, were thus typical in this respect.

The publication of articles and books in Europe about the brain-drain died down after the 1970s. Europe and the United States today form a common market for academics, where it is natural to move back and forth. Today the discussion about brain drain instead concerns Asia and Latin America.

Drawbacks and future

To claim that this highly competitive system has no drawbacks would be misleading. For instance, mobile scholars tend to have difficulties in raising families, especially when women also have careers. Such drawbacks could put constraints on the introduction of the American system as a whole in Europe. Europeans could challenge the American system, however, by adopting parts of it. Sweden gives an example. Compared with the population, the number of Ph.D.s in Sweden lagged for decades after the American rise in production. Yet rather suddenly, in the 1990s the Swedish per-capita production of Ph.D.s overtook the American, see figure 4. However, the part of the diagram covering the period from the 1960s to the 1980s is the most important for the Dovring biography.

Part of the explanation for the Swedish acceleration is a conscious venture from the Swedish government to raise the number of completed Ph.D.s. Another explanation is the decreasing investment in higher education that has been made in the U.S. since the 1980s.

280 Schuster (2002), p. 1 541.

281 Chorafas (1968), p. 23–28.

282 Chorafas (1968), p. 24–25.

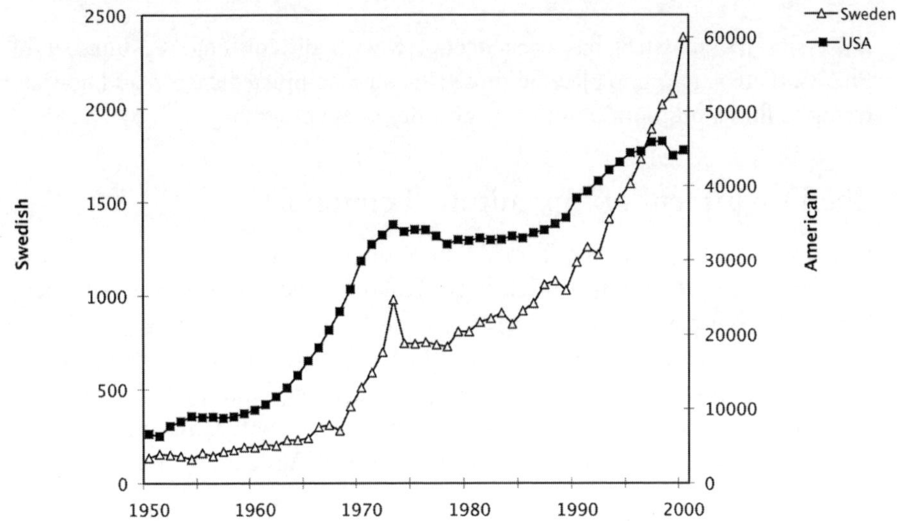

Figure 4. Production of Ph.D.s in absolute number in Sweden and in the U.S. The middle of the diagram, from the 1960s to the 1980s are most important for the Dovring biography.

The Swedish numbers are given on the y-axis to the left, the American on the y-axis to the right. In the beginning of the period, the American population was about 20 times bigger than the Swedish; in the end of the period 30 times bigger, so the scales on the y-axis are related in a factor of 25 (the scale to the left being a 25th-part of the scale to the right).

Source: The diagram from Myrdal and Söderlind 2003, summarizing national statistics: "U.S. Department of Education. Digest of Education Statistics" for the U.S., and "Statistiska Centralbyrån (SCB)" for Sweden.

In figure 4 the number of dissertations are given in absolute numbers, but to make the comparison between Sweden and the U.S. more representative, the scales have been adjusted according to the total population in the two countries. In the first part of the curve it is very clear that even as the Swedish production of Ph.D.s increased in the 1960s, the American production increased faster. This was when the American dominance of the Academic world system was established. This was also the period when Folke Dovring was recruited to the U.S.

The interpretation of the later part of both these curves goes beyond the scope of my investigation, but the dramatic change is not unique to Sweden. Partly, it describes a delayed expansion in Europe. One could also ponder about a loss of steam in the American system, where more post graduate students are recruited from abroad.[283] It should also be mentioned that the state-controlled Swedish

283 The proportion of doctorates earned by U.S. citizens fell from about 88 % in 1977 to around 78 % in 1989, Schuster (1992), p. 1541.

university system lately has been occupied with diminishing the number of Ph.D.s, thereby trying to place all those who have completed degree, and now are trying to find a position, in different post doctorate programs.

The Department of Agricultural Economics

I am going further down the "funnel", from the national level, to the university- and department-level until I reach Folke Dovring. The micro level is explained, but can I also explain the macro-level.

Dovring was hired as a full professor in 1960 at the Department of Agricultural Economics, at the College of Agriculture at the University of Illinois in Urbana-Champaign. Today the department has merged with Consumer Sciences Economics into ACE, which is the Department of Agriculture and Consumer Economics.

The University of Illinois was first established as a typical land grant university in the Midwest, with practical knowledge in agriculture and engineering combined with research and education in the humanities, law, etc. The university gradually strengthened its position during the ensuing decades. I am not going to give the history of the whole university, but will concentrate on the Department of Agricultural Economics.

The diversity of American universities makes any generalization complicated, and not all of the departments of agricultural economics in the Midwest were inbred, mainly doing research about their own region. Nevertheless, dynamic scholars with new ideas tended to meet with difficulties, and sooner or later moved on to other positions. In the 1950s, these departments at the large land grant universities in the Midwest had stagnated and were in danger of losing prestige.[284] Then came a change: from the autocratic regimes with inbred faculties, to open and international departments with a growing faculty influence.

Willard Cochrane, in his interesting history of the Department of Agricultural Economics in Minnesota, also has given a general analysis of how many of the Midwestern universities from about 1930 to 1950 were dominated by what he calls "the great man phenomena". One powerful head led the department in a conservative fashion, focusing on work in farm management and marketing, often serving the local need perfectly but seldom reaching beyond. There were exceptions, such as John D. Black of Minnesota who went to Harvard in 1928 or T.W. Schultz of Iowa who went to Chicago in 1943 (and later was awarded the Nobel prize), but, significantly they both had difficulties before they left.

284 To describe the change I have, besides interviews, mainly relied on Willard Cochrane's analysis of the history of Agricultural Economics at the University of Minnesota (Cochrane 1983), Franklin Reiss more descriptive history of Department of Agricultural Economics at Urbana-Champaign (Reiss 1982) and Harold Halcrow's auto-biography from 1998 (Halcrow 1998).

In Minnesota the department, from late 1920s to the mid 1950s, was steered with firm hand by the head, Oscar B. Jesness. During what Cochrane labels the Jesness era, "for faculty members not in tune with Jesness philosophy, life could be extremely difficult".[285] Cochrane affirms that such a leadership would have resulted in revolution among the faculty had it been in the 1960s and 1970s.

This regime established parameters for the faculty. The very same descriptive method was used in every investigation, and theory played virtually no role. "Economic theory played little or no role in guiding and directing research" and a staff member "could be reprimanded for hiding in the library".[286] Give the producers the facts, was the philosophy. All vacancies were filled with Ph.D.s from Minnesota, and the only one who had not received his degree at Minnesota "was ill-treated professionally by his superior, O.B. Jesness". Cochrane concludes that "the professional staff of the Department by 1950 was badly inbred".[287] Courses for students did not change; indeed: "Change for whatever reason, was not popular in agricultural economics at the University of Minnesota in the Jesness era."[288] During the depression years, the department could not defend its position, and got fewer resources.

After 1950, Jesness started to change strategy and hired people from other universities. Cochrane himself was one, Philip Raup (Dovring's friend) another. Jesness once commented on this to Cochrane, and his only explanation was that: "It seemed like a good idea at the time".[289] Apparently, the old head was clever enough to see a new time coming. Even though Cochrane was hired by Jesness, his judgment over the regime is harsh, and he ends his description of this gloomy period in the history of the department with a pledge for the freedom of thought and research: "Subject matter departments in a university must be concerned with problem solving, exploring the unknown, and idea discovery. This kind of research endeavor takes time and is subject to failure."[290] Administrators must recognize this.

After Jesness left in 1957, the department totally changed. Faculty was hired from all over the States, and, instead of only working in-state, the faculty started to work on the international arena.

A similar change occurred at many departments of agricultural economics in the Midwest in the 1950s. One direct reason was an increasing need for research in the field and a growing amount of resources available. However on a more general level, this was a part of the new paradigm established after World War II.

285 Cochrane (1983), p. 27.

286 Cochrane (1983), p. 30, 40.

287 Cochrane (1983), p. 34, 32.

288 Cochrane (1983), p. 36.

289 Cochrane (1983), p. 34.

290 Cochrane (1983), p. 40.

Illinois is another illustrative example. The department was founded 1932, and H.C.M. Case had been the head from 1934 to 1955. He certainly did contribute important work, not the least in the international field, but the department had under his regime become heavily inbred. I am not going to write the detailed history of the department, but Case was according to my sources, a very kind but quite conservative leader of the department.

A list of tenured faculty or faculty on the tenure track from 1930 to 1975 shows that the change started at Illinois at the same time as in Minnesota, just after 1950, even before Case retired, see table 2. Yet notably, it was with a new department-head in 1957 that a total change was instigated.

Table 2. Academic personnel with Ph.D.s at the Department of Agricultural Economics, Univ. of Illinois, Urbana-Champaign, with rank of assistant professor and higher.

Start of service	*Ph.D. Ill.*	*Ph.D. other univ.*	*B.S. abroad*
1930–4	3	1	0
1935–9	2	3	0
1940–4	4	0	0
1945–9	10	2	0
1950–4	6	5	0
1955–9	2	5	1
1960–4	0	11	1
1965–9	4	10	3
1970–4	2	6	0
1975–9	3	13	1

Source: Reiss (1982).
B.S. = Bachelor of science. Ph.D. Ill. = doctoral degree in Illinois, nearly all of whom from this department.

In the whole College of Agriculture at the university the number of foreign students rose dramatically during these years. Before the 1940s there were nearly no foreign students taking a Ph.D. degree, in the 1960s they were a quarter to half of the Ph.D.s.

Table 3. Ph.D. degree – total number and foreign students 1945–1969.

Year	*Total Ph.D.s*	*Foreign students*
1945	4	1
1946	3	0
1947	9	1
1948	11	1
1949	12	3
1950	17	5
1951	23	4
1952	29	5
1953	20	4
1954	36	7
1955	54	12
1956	51	12
1957	28	10
1958	39	10
1959	36	13
1960	56	23
1961	45	10
1962	50	17
1963	50	9
1964	35	9
1965	44	15
1966	48	24
1967	48	22
1968	54	28
1969	74	24

Source: Moores (1970), p. 239. Among those taking the M.A./M.S. degree the foreign students were much fewer.

Harold Halcrow, who was hired to change the department in 1957, has in his autobiography and in an interview described how this change was carried out.[291] I have also spoken with other retired professors still living.

The University of Illinois had some years earlier gone through a major crisis. A new president, George D. Stoddard appointed in 1947, had tried from above to force through a modernization of the university. He was a psychologist by

291 Halcrow (1998), and interview with Harold Halcrow, 24.5.2002. I have also presented this in a lecture at the ACE in the fall of 2002, and Harold Halcrow then generally confirmed my interpretation, even if it must stand as mine.

training and belonged to the radicals in the New Deal-era. He chose new deans to usher in reform and got involved in a series of conflicts, mainly with more conservative groups at the university. Intense strife went on at the Department of Economics, having both personal and ideological overtones, but the Department of Agricultural Economics was not directly affected by this conflict. Later, several of the newly appointed professors in economics left the university and went to other universities. Some of them made quite reputable careers (one of them, Frank Modigliani who had been on the president's side, actually received the Nobel Prize). In addition, not surprisingly, some of the more conservative "old-timers" also left.

Eventually, the president had to resign his office and leave the university in 1953. (He went to New York.) Many still remember when he came out on the balcony to his house, and explained to the crowd that had assembled there why he had decided to leave. This conflict was a national disgrace for the University of Illinois in the academic world, but it can also be seen as one of the battles fought as the new paradigm forced its way through.[292]

The dean of the College of Agriculture, Louis Howard, and the new head of the Department of Agriculture Economics, Harold Halcrow, had this failure in mind, even though they did not speak directly about it. The mistake the earlier president and his group had made was that they had tried to move too fast, so it backfired. Halcrow's strategy was to build confidence with the older professors, but not tell them directly that he planned to change everything. Still, he was very clear in the interview, when he explained to me: a leader of a group must never tell a lie. He held seminars and used much face-to-face communication to transmit his general ideas before the change started and during the process.

A problem was that he could not hire good young faculty without being able to offer them good salaries. The example from the big battle at the university some years before had shown that it was not a good idea to give the new faculty much higher wages than the old faculty. (To some extent this will always be a problem, and it even has a label: "compression".) Thus, he obtained a considerable sum from the dean and raised the wages for the best of the faculty already in the department. Then he was able to offer this new higher salary to candidates he wanted to recruit from other universities.

He also quite aggressively tried to get tenured faculty from other departments of agricultural economics, which caused some grief among his fellow department-heads in the Midwest. Paradoxically, as the whole system started to change, other more aggressive heads also threatened his faculty. To counter, Halcrow made

292 The literature about this interesting conflict is unfortunately rather meager; see Brichford (1995 mimeographed 4 p.) and based on this mimeographed text Conley (1998), p. 509. Concerning the conflict in the Department of Economics I rely on the personal memories by several of those retired professors that were young students at that time, and who were discussing this with me several times.

agreements with several other heads that he would not try to steal their best faculty, if they would tip him about their best graduate students, so he could hire them when they graduated.

We can now see how the new system takes form on a local level. After a while, these secret, but formalized, agreements were no longer necessary. They became internalized in the system – a system where it became natural for a professor to try to send away his best students instead of keeping them. To keep them would have been to hamper them, as the culture had been changed. Someone who did not test his ability in another milieu was regarded as inferior.

In such a climate, it was also natural for Halcrow to extend his search abroad. As he writes in his autobiography, when Charles Stewart, the professor who taught land economics, left they wanted "the most outstanding person in land and resource economics they could find". The offer went to Philip Raup at Minnesota, who at first was interested but then declined. He wanted to stay in Minnesota, which he actually did for the rest of his career. Instead Raup suggested to Halcrow: "If you want to hire the most able person in the world who is working in this area, you should go after Dr Folke Dovring."[293]

Philip Raup had worked in Europe and knew of Dovring (they had met in the FAO-library in Rome). Halcrow immediately borrowed *Land and Labor* from the excellent library at Urbana-Champaign, and was impressed. After he had acquired letters of recommendation, he sent a letter to Folke Dovring, who at that time happened to be in the U.S. And as we already know: Dovring accepted the offer.

The Department and Dovring as a part of the leap forward

The changed direction also had an immediate effect on the quality of the academic endeavor at the Department of Agricultural Economics at the University of Illinois. An important proof of this is the number of theses presented, see figure 5.

In the 1920s, only masters were presented at the department; the first Ph.D. was finished in 1931. Thereafter, in the 1930s, occasionally Ph.D.s were presented. The World War II caused a dip in the production here as elsewhere in the nation, but was followed by production of Ph.D. theses on a grand scale. After having reached a plateau around 1960, there was a momentum increase in the late 1960s and the 1970s. A stagnating 1980s was followed by a new but relatively smaller increase at the end of the century. This mirrors the national change.[294]

---

293 Halcrow (1998), p. 173–175, interview with Halcrow 24.5.2002.

294 Another example is the corresponding department at University of Minnesota, where the change was very similar to the one at the Department of Agricultural Economics in University of Illinois. See Cochrane (1983), p. 83–108 and Sundquist (2001), p. 36–56. The number of Ph.D.s rose from around 2–3 in the 1930s, to around 6 in the 1950s, 7–8 in the 1960s–1970s, and a further increase to around 10 in the 1980s–1990s, with dips

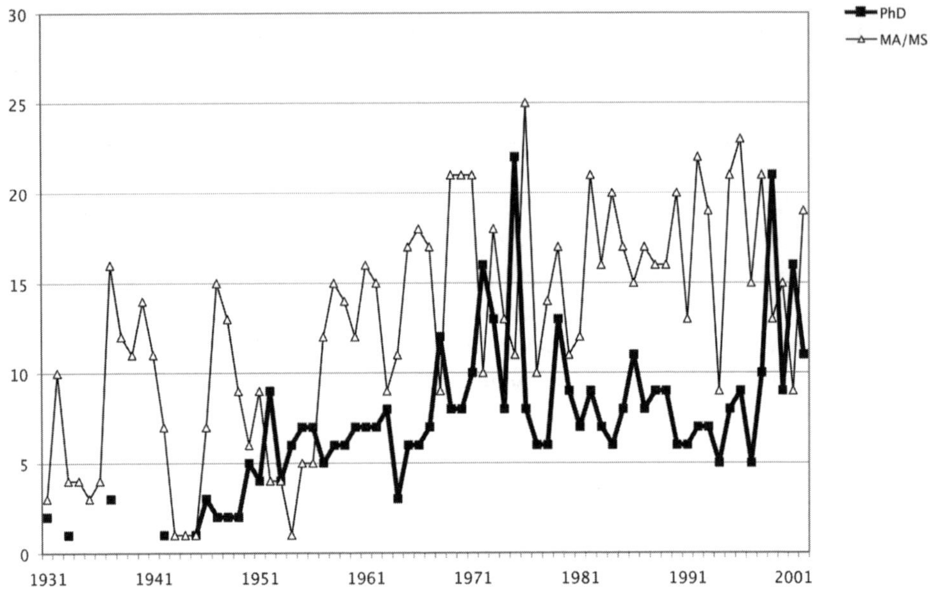

Figure 5. The number of Ph.D. and M.A./M.S. theses at the Department of Agricultural Economics (including later associated Department of Consumer Sciences) and today's Department of Agriculture and Consumer economics.

The Masters appear above the Ph.D.s; over the period the total number of masters is nearly double the number of Ph.D.s. (Note that most masters did not present a thesis, and are thus not included in the diagram.)

Source: Investigation by Andrew Isserman, who has been kind enough to give this to me.

This leap forward did not come automatically but was achieved by the hard work of individual professors and graduate students, both on the national level and on the department level. Here I will single out Folke Dovring as one of the local heroes without whom this success would not have been possible.

During his years as a professor, from 1960 to 1986, he was the main adviser to 24 Ph.D.s and also to 37 master's theses. In supervising about one tenth of the total number of doctoral graduates at the department, he was second only to Chet B. Baker (who produced the astonishing number of 47 Ph.D.s). Folke Dovring and his friend Earl Swanson were nearly equal in their production of degrees at the department. These three professors, who served approximately during the same crucial period of 1955–1985, contributed with more than a third of the doctoral dissertations. Other professors certainly made contributions during this leap forward when teaching for undergraduates also increased tremendously. We

during World War II and also around 1960 and 1980. The number of MA/MS also followed this curve.

should also not forget the "extension", informing farmers, which continued to be important.

Concentrating on Dovring's contribution I will mainly talk about the graduate teaching, but he also taught at the undergraduate level. His main course in graduate teaching was in land economics. Students were impressed by his scholarly breadth. In a preserved evaluation from 1971, one of the students states that Dovring is "one of the most knowledgeable professors in this area that I've ever taken a class from", and another pronounces him "extremely knowledgeable in his subject area". Students nevertheless complained that he was not always easy to understand.[295]

Dovring enjoyed mentoring graduate students. Even in the late 1940s, when he received his first appointment as an assistant professor in Lund, he wrote to his mother about looking forward to supervising students and molding their scientific education and conception of truth.[296]

At Illinois he often took on the graduate students no one else at the department wanted to supervise; he tended to attract students from non-traditional backgrounds and foreign students. When they had difficulty writing English, he often told them to compose a draft in their native language. Because of his own linguistic skills, there was a likelihood he could read it. The majority of his students wrote about the Third World, often countries from where they came. None of the dissertations focused on historical subject matter, but some were about land reform with a historical introduction.

That Dovring took on, or was chosen by, non-mainstream students did not mean that they were let through easily. In fact, he was considered demanding. For instance, in a preserved letter to the head of the department about a graduate student who wanted to proceed from a M.A. to a doctoral thesis, Dovring stated that the student had not worked hard enough.[297]

Especially when he had chosen the topic himself, Dovring would maintain firm control over how the work was proceeding. One graduate student, David Chicoine, recalls that he was handed material that Dovring had collected (about land prices outside Chicago), and they met at least once a week.[298] One of Dovring's research assistants, Tim Bloomqvist, relates that students always had to be well prepared when Dovring arrived, most often once a week, since he wanted to know what had been accomplished during the preceding week and what was slated for the next.[299]

Half-a-dozen of his students became professors at different levels in the U.S.

295 DA, Course and Instructor Evaluation, Fall Semester 1971, p. 10.

296 PA-MS, Folke Dovring to Naemi Ossiannilsson, 31.10.1948.

297 DA, Examination of V.V. Sharma, 27.8.1965.

298 Interview with David Chicoine, 18.6.2002.

299 Interview with Tim Bloomkvist, 17.6.2002.

and some secured positions in their home countries. Two of them attained positions as heads of their departments.[300]

Serving others with information

Dovring was not much of a casual talker in the hallway, and did not attend coffee-breaks together with other professors. Most of the time, he sat in his small room or in the library. According to his head, Halcrow, when the department was rebuilt, and subsequently rooms were distributed, Dovring had chosen the smallest room available to forestall that he should ever be pressed to share it with someone.

Meetings between Dovring and graduate students or others, who came to visit him in his room, tended to turn into freewheeling intellectual discussions where one topic led naturally to another. Dovring might remove a book or an article from the bookshelf and ask, "Have you seen this?" A discussion with Dovring often took an unexpected turn. Earl Kellogg, associate provost for international affairs at the University, was then a young professor. He found that when he emerged from Dovring's room after a conversation, it was a struggle to come to grips with what had been covered, but when starting to understand felt that he had learned something.[301] One of Dovring's most successful and beloved graduate students, David Chicoine, reports that he always turned last to his supervisor Dovring for advice. He knew that Dovring would suggest something totally different and unexpected, whereas others would merely confirm what Chicoine himself had already considered.[302]

During the work with this book, I happened to listen to Stephen Bunker from the University of Wisconsin at a conference (about a totally different subject). Bunker mentioned Dovring in his speech. Talking with him afterwards, he related that, during his stay at University of Illinois from 1978–1980, Dovring had meant quite a lot to him in his research about Latin America.[303]

Dovring's language skills set a positive example for younger colleagues, one much needed in a largely monolingual country like the U.S., where language isolationism reduces the possibility of understanding other cultures and mentalities. A then junior-faculty member, Randall Westergren, told me that without advice from Dovring he would never have learned another language; now he could even teach in French.[304]

---

300 This is based on a search in Google for all of the doctorates who graduated with Dovring as main supervisor. The two who reached positions as head of a department or even further up in the academic hierarchy were Herbert Stoevener and David Chicoine, who later also became one of the vice-presidents at University of Illinois.

301 Interview with Earl Kellog, 11.7.2002.

302 Interview with David Chicoine, 18.6.2002.

303 Interview with Stephen Bunker, 21.5.2002.

304 Interview with Randall Westergren, 20.6.2002.

A duty Dovring avoided was administration committees. He, however, served loyally on other committees, as the one for East European studies.[305] He did not organize many seminars himself, even though one of his first duties was to sit on a committee to organize a seminar-series about methods for graduate students.[306] The anthropologist Julian Stewart was the most famous to be invited to the series, but other professors from the University of Illinois were also invited to give lectures. Among these were the agricultural historian Fred Shannon, who wrote about frontier farmers, and the sociologist Louis Schneider. Dovring also managed to put Karin Dovring on the list. Dovring, however, does not seem to have collaborated with these scholars later during his stay at the University.

As a scholar he did not often work together with others, and he very seldom published jointly with his graduate students. This was in an old-fashioned humanities-tradition.

The European scholar in the U.S.

Taken all together, Folke Dovring was one of the professors who served one of America's top universities to accomplish the great leap forward in the 1960s and the decades that followed. He supervised many in the wave of new doctorates that the university produced, and, typically, many of them were from foreign countries. He helped, but to a lesser degree, to take care of all the new undergraduates that swarmed the campus. In addition he helped and advised young scholars in the beginning of their careers. Many at the university considered him to be among the most erudite members of the faculty, someone who provided a fascinating answer to virtually any question.

Many of these new approaches Dovring gleaned from his wide reading in varied subjects. Apparently, he did not feel the need to travel to conduct research, as he had done in his younger years, but was satisfied by excursions to the extensive collection in the University of Illinois' library, which has one of the largest collections in the U.S. of scholarly literature, in many languages. In addition to Swedish and foreign languages studied as a young boy – Latin, Greek, English, German, and French – he later became familiar with Spanish, Italian and Russian. Through these, a reading knowledge of additional European languages was acquired. (Late in life he started to learn Mandarin.) When students and faculty met Dovring casually, even many years after his retirement, it was in the library or walking to and from the library.

It also seems that Dovring tended to become more European the longer he lived in the U.S. He became the European gentleman at the Midwestern university, walking across campus, properly-dressed in suit and vest, using his umbrella as a walking stick. Of course he would never have been so "European" had he

305 Interview with Ralph Fischer, 23.5.2002.

306 DA, Working List for Agricultural Economics Staff and Graduate Students, 15.6.1960.

stayed in the old country, as Europe tended to change, and in an Americanized direction.

This is only the superficial view of the old-fashioned European in the frontier land, a country where the slogan (from Dovring's assumed standpoint) seemed to be: if anything is proper, we don't. It is nearly a caricature, and of course Dovring, in many respects over the years, was Americanized. More interesting is to understand the role he strived for and acquired in the country where he readily sought and acquired citizenship. (At the time, this meant giving up his Swedish citizenship.)

To summarize, Dovring was excluded from Europe and from an academic society that had similarities with what we have met in the Midwest in the 1950s, as just described. In the U.S., however, the leaders of the university system, with the gradually growing support from the faculty, broke away from this self-absorbed system. The reasons for the accomplishment of this more open-minded system are complex and rooted in earlier history. One result was that a leading land grant university in the heart-land of the U.S.A., around 1960, was able to offer Dovring the freedom of thought that he was seeking. This was the main attraction – the "pull factor" – that caused him come to the U.S.A.

The university system in the U.S. was furthered by recruiting hard-working and knowledgeable scholars such as Dovring. They came en masse in a "brain fill", and they helped the big country to take the leap forward in higher education, and thus to become the leading nation in the university system of the world. This has significance to the knowledge based economy of today's world.

The next and crucial question to be answered is: to which fields did Dovring direct his research? Advising and teaching is an important part of a professor's work, and developing new ideas and new knowledge is another. How did he use the time for research and freedom for which he had been striving for so many years, and which he now had? And what became the reception of his ideas in the new country?

CHAPTER 6
Economist in America (1960–1987)

A free scholar

Dovring wanted to live a good life of hard and productive scientific work, do research he had chosen himself and use methods of his own. When he became a full professor at a well-known university in the U.S., this dream seemed to have become true. He was 44 years old. He and his wife had spent ten years in a rootless existence, in Switzerland and in Italy, with Karin often traveling to the U.S. (to Yale). They had left Sweden, but had not really settled anywhere else. Now they had found a home; they had bought a wonderful castle-like and Italian-styled house in Urbana. And he had a good position.

Formally, he replaced the expert in land economics; Charles Steward. Over the years, Dovring gradually came to present himself more often as a professor of "Land economics" than of "Agricultural economics". He probably could have lived out his life coping with this sector of the economy, being the expert in land-economics at the Department of Agricultural Economics at the University of Illinois. It would have been a respectable choice, and a fine assignment. But Dovring had another agenda.

His first five years was devoted to following up *Land and Labor* (see foregoing chapter). Thereafter, he gradually left historical questions, and, when he became an American citizen in 1968, he started to formulate a reform program for American society. During the last fifteen years before his retirement, this was foremost on his mind.

To accomplish this, he required intellectual freedom. Unfortunately, immediately upon coming to the department, he became involved in a project he had not chosen and disliked. His head of department at that time, Harold Halcrow, told me what happened: "Charles Stewart, his predecessor, had gotten funding to make a report about the recompensating to the farmers from the highway program. Eisenhower had decided we should have a highway program, if the army had to be transported fast from one end of the country to the other. Stewart did not know how to do it. I asked Folke, and Folke said 'I think I will do that'. I said that 'farm management could help you'. And Folke got the job done. That was Folke."[307] This was an Illinois State project, but also a part of the total rebuilding of the U.S. transport system after the World War II. The whole country was in-

307 Interview with Harold Halcrow, 24.5.2002.

terconnected with a web of interstate freeways.

Apparently he, already at that time, had started to formulate ideas about the car-oil complex as destructive and wasteful. Folke disliked the project so much that he had his name put on the report only as "project supervisor", even though he had written most of the text and also had drawn the maps.[308] He disliked being forced to finish what other scholars had started. Even more, he was against the whole idea of promoting motor traffic and wasting good farmland to make freeways.[309] However, he figured out a method to pay the farmers, by measuring the loss that individual farmers incurred when their landed property was cut into pieces. He collected a huge amount of statistics on landed property in the state, and several times later came back to this material.

It was an irony that he, who was against the car-complex, was one of those who made a major contribution to the expanding of this complex. Halcrow realized that Dovring did not like the task given to him and never again asked Dovring to do anything specific.

By carrying out his first task successfully though and at the same time expressing his dissatisfaction with it, Dovring managed to establish a certain amount of freedom for himself. His department-heads afterwards did not know the exact nature of his research, but they were convinced that whatever he did, he did it with excellence.

Economy is a normative science, not only researching the society but also suggesting amendments. When Dovring elaborated a reform program, and took a standpoint on the society and economy of the U.S., he did not come out as a mainstream economist. One aspect was that he continued to advocate support for the small farmers against large landowners. He thus also recommended land reform in low-income countries, even if that meant a rather harsh treatment of the ruling class, which often was the group working together with the U.S. government. Specifically, the basic idea for his reform program was the identification of waste as the crucial problem in our society, especially in the U.S. He railed against waste of oil, and thus also against cars and actually against the whole logistic structure built up in the U.S. Other out-of-the-mainstream ideas were that he believed in a more egalitarian society, in more state-control and in restrictions against the full force of the free market.

In many respects, he stood on the side of the liberals in the U.S. He joined the Democratic Party. He drew a line against the far left and condemned the Soviet-system as centralized and undemocratic. Basically, he supported the capitalist system. He also looked upon the U.S. as a country with a mission in world history, to spread democracy and freedom, and he was proud of being an American citizen. His whole program for change was rooted in this feeling.

308 Dovring (1965).

309 Later he argued that too much good farmland had been wasted to build shopping centers, highways, etc, see Dovring et. al. (1982).

Professor Dovring at his home in Urbana, in the 1970s.

Reaching out?

Dovring's time in America can be divided into three periods. The first ten years, he worked mainly with historical questions and with questions connected with land reform in the third world. The following fifteen years, he worked with research connected with his reform program publishing the results in articles. The last years, from around 1985, he published a series of books on his reform program.

When he wrote about history, he did publish in prestigious journals. However, the more he committed to doing research about questions connected with his reform program, the less success he had in spreading his publications. Actually, many of his publications after the mid-1970s were mimeographed department papers or articles in local journals. Perhaps his articles were not accepted in more prestigious journals as he did not like to adjust his writings to what publishers and reviewers suggested. I have no direct evidence on rejected articles. He did continue, however, to publish especially reviews in renowned journals such as *Land Economics*.

Dovring, in the 1980s, told his favorite graduate student, David Chicoine, that he considered the peer-review system to be merely a form of censorship.[310] By necessity, every kind of selection involves some version of censorship, and Dovring had, in earlier years, worked smoothly with this system. Actually, for many years, he peer-reviewed for *The Journal of Economic History*, and he sat on the board. Indeed, his sometimes rather harsh reviews of other authors indicate that he himself executed some "censorship".

My assumption is that Dovring deliberately stayed out of the game. The advantage of such a position was that he could think and speak out more freely; the disadvantage was that he did not reach out. He wanted to reach out. That was why he worked so hard with book writing at the end of his life. Nevertheless, these later books are generally not mentioned in other books, and seldom in the citation index over articles.

Similarly, his later articles and books are not mentioned in textbooks, and seldom in the citation index. They were reviewed in some prestigious journals, but the reviews were few and critical of the content of his program and the scientific value of his books. Though he saved many reviews on earlier books in his archive, none on his later books was saved. The reason could be that he did not receive them from the publisher, but perhaps also that he disliked them.

He felt that the reason why his reform program did not catch public attention was that his ideas were premature. Probably, he felt like a prophet from the Old Testament; a prophet coming down from the mountains to tell what he had realized about the wrath of destiny whose his voice was only heard as crying in the wilderness. In a letter to his friend Philip Raup he wrote, late in life:[311]

> "The trouble is that most people are interested in what happens immediately, in the very the short run, but not how we might steer events on the longer haul. The current crisis could have been averted by a national energy policy started ten years ago, just as the Viet Nam War could have been avoided by land reform in the late 1950's. We both know how that was handled – like 1980's economic policy, wishful thinking was allowed to pass for conservatism. So when are we going to mind the future?"

Here he blames politicians and, as a consequence, also the American academic society as to why so few cared about what he wrote. But was it that simple? Could it be his own fault that he did not reach out?

310 Interview with David Chicoine, 18.6.2002.

311 PA-PR, Folke Dovring to Philip Raup, 19.12.1990.

Land reform, 1960–1975

The small family farm had been on his agenda since the 1950s, when he wrote *Land and Labor*. When he turned his interest to non-industrialized countries, his focus became the need for land reform. He saw land reform as a way to solve problems with surplus population. Even in the early 1960s, he stated that rapid reform was needed around the world but reactionary regimes wanted to delay it.[312] His main argument was the economic efficiency of the small farm. Furthermore, he considered it a false argument that social justice had to be paid with economic decline.[313] Undeveloped countries had an abundance of labor needing employment.[314]

His own research contribution was summarized in a couple of mimeographed reports for the U.S. state organization AID (Agency for International Development), published at a conference in 1970. Mexico was his major example. His investigations had been published some years earlier, first in English and then also in Spanish, and the results attracted some attention, not the least in Mexico. He carried out different statistical procedures to prove that the Mexican ejidos, cooperative communes for small farmers, were more efficient than the large estates. During the long period 1930–1960, the eijdos had had a relatively impressive increase of production by international comparison. Dovring's conclusion was that the ongoing land reform, especially after president Cárdenas in the 1940s, aimed more of the nation's resources at labor-intensive growth, which lead to land clearances and irrigation.

In other special reports he treated Hungary and Yugoslavia. He was generally critical of the collective farming system imported from the Soviet Union, which he saw as a parallel to the big agro-business in the West. He also wrote the summary for the whole project on the economic effects of land reform, with examples of rather successful reforms in countries, such as North Vietnam and Iran. He concluded that the reason why land reform promoted, rather than hampered development, was that smallholding could substitute labor for capital.[315] Low-income countries needed greater equity and more education rather than labor-saving techniques, and land reform was a way to instigate investments in human resources.

Under specific circumstances, land reform could be a failure, at least temporarily. For instance in Kenya and Algeria, Europeans managed agriculture better

312 Dovring (1962b), p. 35.

313 Dovring (1964), p. 95.

314 Dovring (1962b), p. 30.

315 In a study about India, from 1972, he again came to the conclusion that low levels of income meant that the population had a shortage of everything without labor-force, and thus labor-intensive production was to be preferred and land reform could distribute land to suit such a development.

than natives. Therefore, land reform had caused problems in these countries. Such an opinion was not in tune with the anti-colonial movement, which at this very time was spreading across the world.

In 1974, he summarized his ideas at a conference in Germany, where he quoted his earlier papers, and argued that there had been a shift in informed opinion about land reform as a result of the conference of 1970, referred to above. He states that the belief in large estates as economically more efficient is false, and declared that land reform would have been "the best antidote against communism" in South Vietnam.

Dovring was a part of a more general intellectual creed regarding land reform. Redistributing land to those who work it, has a history clear back to the Middle Ages until the present, but after World War II it became a burning issue. Peasants all over the world emerged as important actors in history; in the European partisan-movements as in the Asian guerilla wars. Land reform was a top priority for the newly-founded United Nations. Indeed, in three bibliographies on land reform and related issues (published by the U.N. organization FAO in 1952, 1959 and 1972), more than ten thousand publications are listed.[316]

The United States, at the outset of the Cold War, began to look for ways to meet the communist challenge, for instance instigating land reform in Japan. Actually, land reform well-suited the nation's international assignment. President Truman declared that he believed in land reform in Asia, which would promote the family farm and democracy.[317] Conversely, other Americans' interests, particularly in business and in power politics, gradually came to see land reform as a subversive threat. Socialists particularly took up the issue. Land reform also was a threat to large landowners in political control in countries that were close allies to the U.S. Dovring later pointed at 1952 as the decisive moment when anti-communism in the U.S. lead to a dissociation from land reform.[318]

From an American point of view, the importance of the 1970 conference was significant. It was important to prove that land reform lead to economic development, in order to arrest a diminishing political support for this movement. In a worldwide survey over the question some years later, in 1977, the English economic geographer Russell King championed much the same ideas as Dovring, and several times quoted Dovring – both his study from 1970 and *Land and Labor*. The main idea was that land reform affected the economy in a positive way. Dovring also reviewed King's book favorably in *Economic Development and Cultural Change*, but as always with reservations. He remarked on two weaknesses: that King was not critical enough of communist China, and that King did not realize the deep cultural difference between eastern Asia and southern Asia.

Another expert on this issue was Peter Dorner, who, in his textbook from

316 King (1977), p. 3.

317 King (1977), p. 44–45.

318 Dovring (1987b), p. 410.

1971, has many references to Dovring. Again it is his survey of Mexican land reform which is quoted. Dovring's emphasis on small scale and labor intensive agriculture as a way forward is cited with approval. When Dorner, twenty years later, summarized research on land reform, he again referred to Dovring's Mexican articles.[319]

More left-wing scholars had a partly different opinion. For instance, his old colleague and friend from Rome, Erich Jacoby, published a worldwide summary in 1971. He had been the head of the Land Reform Branch at FAO when Dovring worked there, and later he was recruited to Gunnar Myrdal's institute in Stockholm – occupying a similar position to the one that had been earmarked for Dovring some years earlier. Jacoby did not quote Dovring, and Dovring did not mention Jacoby, but King quoted both of them.

Jacoby is more keen on the social aspects of land reform than Dovring even though he, as Dovring and King, did emphasize the need for labor-intensive technique. For instance, Jacoby is less optimistic about the Mexican example, following the Marxist sociologist Rudolf Stavenhagen, who emphasized the fate of the poor and land-less.[320] On the other hand, Jacoby is more favorable to the socialist reforms in China. His opinion is that, even if land reform is not immediately economically efficient, it is the prerequisite for other development as it changes the social structure. Dovring preferred to speak specifically about the economic effects.

The world movement for land reform retreated after the 1960s. At first it seemed that all that this literature on land reform amounted to was dreams soon to become crushed. Indeed much of what happened was irreversible. The zamindars in India lost their position, the eijdos in Mexico played an important role for decades, and in Chile the land reform was stopped, but not rolled back, after the coup in 1973. There are no total victories in history.

Some decades after World War II, the small farmers around the world advanced their positions, giving an impetus to economic development. The family farm gradually has become dominant in larger parts of the world, even though not always in its Western form.

Dovring was one of the scholars who understood this and gave the theoretical tools to analyze and defend the process. He, at this time, remained in the middle of the discussion,[321] and, in the 1970s, he was considered as one of the leading non-marxist experts on land-reform.

---

319 Dorner (1972), p. 46–48, 95, 111; Dorner (1992), p. 25, 88. Ronny Pettersson made me aware of Dorner's quotations.

320 A strange fact is that when Dovring proves economic success for land reform in Mexico, Jacoby (1971) p. 81–82, 346 talks about stagnation and failure.

321 In Dovring (1988b) about food production he came back to the problem and again pointed at successful examples, pp. 105, 242–244.

Small American family farms, 1962–1970

The distinction between small or medium scale farming and family farming is blurred. In the last decades we have had a very strong concentration of farms in Europe (even more in the U.S.), which has not changed the basic structure of the farms as work-units. Members of one household often manage even large factory-like units, with thousands of hens or pigs or with thousands of acres of land. The workforce can consist of a father and son or of a husband and wife, with perhaps one or two farm-hands, but seldom more.[322]

The reason why economies of scale seldom go over a two-man farm has historical, cultural, economical and ecological roots. The family farm was the basic unit in European farming. It became even more so with the dissolving of feudal or semi-feudal systems, where large estates with many subordinate workers had played an important role (even if they seldom held more than a part of the total area under direct cultivation). In the United States, the family farm became a part of the national ideology, both as a result of emigrants opposed to the feudal structure of European agriculture, and as a result of the victory of The Union in the Civil War.[323] To champion family farms was a part of the democratic foundation of the society.

However, cultural and historical factors only furnish a part of the explanation. The handling of living material, crops and livestock, is better executed with direct commitment and even dedication from the workforce, which gives the household unit, as a firm, an advantage over larger units. This has, with full evidence, been proved by the giant Soviet experiment, to which Dovring continuously called attention.

When Dovring studied European agriculture during the nineteenth and the early twentieth century, his standpoint was a defense of the medium-sized family farm. He did not defend it because of intrinsic values, but because he considered it more efficient in the use of labor.[324] The family farm relies on the willingness of family members to work for the household, and he could prove there was a strengthening of family farms during the period studied.[325]

Defending the family farm was a theme he followed also in his new home-country. Notably, one of the first articles he wrote was about the Danish farms as a contribution to a debate started by an influential extension-economist at the department of Agricultural Economics, at University of Illinois, Larry Simerl.

To understand the debate, one has to realize the importance of the Danish

322 See books on American agricultural history, as: Gardner (2002), p. 48; Cochrane (1993), p. 360.

323 Gasson & Errington (1993), p. 77.

324 Dovring (1965), p. 143–144.

325 Dovring (1965), p. 189, 376–378. A new legacy of thought developed where proud and influential farmers replaced the earlier suppressed peasants.

example. After breaking the manorial rule around 1800 and the early nineteenth century, agriculture became the motor in the Danish economy. Exporting pork and butter to England, the Danish farmers contributed to the industrialization of Europe. The family farm dominated totally and was organized within a strong cooperative movement, which still holds an important role in Denmark. The Danish example had been under discussion even in Illinois, for instance E.J. Perry's, *Among Danish farmers*, of 1939. So when Larry Simerl, in a wide-spread "extension" letter, sent to many farmers in Illinois in December 1961, declared that Denmark had destroyed the family farm, he was confronting an established knowledge. According to Simerl, the Danish policy of supporting and protecting the family farm was a failure. The size had been kept down, and incomes curtailed. Danish farms did "not present a picture of prosperity". He concluded that government regulations tended to perpetuate farm poverty.[326]

As Illinois was crowded with farmers with a Scandinavian heritage, Simerl got several angry replies from them. Professor Earl Swanson, of Swedish descent and an early friend of Dovring at the department, asked him to write a scientific answer, which would settle the question. Dovring soon published an article together with a Danish visiting scholar, Hans K. Larsen. Referring to Simerl's letter, Dovring and Larsen showed that, in dollars worth of currency, the Danish farmers were on an equal level with that of U.S. farmers.[327] They also pointed at how the well-organized Danish agricultural system, with strong farmers' organizations, could react to and handle new situations in the market – building an efficient chain from the farm to the food industry and to the consumer. In 1962 Dovring followed up with an article celebrating the centennial of the Homestead Act of 1862 in which he heralded European reaction to the Act. His hypothesis was that the act was a part of a parallel ideology in America and Europe, which shaped and reshaped the farm structure during the nineteenth century.

Dovring, in his study about compensating farmers for land used to build the interstate freeways, had collected data on 16 000 real estate sales in Illinois.[328] Later he returned to these data to discuss other issues. For instance, he proved that equity of incomes characterized the U.S., compared with Latin America.[329] Later, in the 1970s, he reflected over a rapid rise of farmland prices, and, in the early 1980s, he made source-critical studies about the size of farmland in Illinois. These proved that official statistics had overestimated the total area by one or two percent. This was the kind of research he could have furthered, had he just been the expert on land economics.

Around 1969, he was uncertain about the future of American agriculture. In the prestigious *American Journal of Agricultural Economics* he suggested two differ-

326 Simerl (1961).

327 Dovring & Larsen (1962c).

328 Dovring (1963).

329 Dovring (1973), p.146–147.

ent futures at the end of the century: family farms or "factories in the fields". He suspected that those who advocate the last-mentioned alternative would win, not because they are right, but because they are believed. Propaganda would play a decisive role.[330] However, he also argued, in his and Karin's *The Optional Society* from 1971, that: "recently has it become clear that 'factories in the fields', the large-scale organization of agricultural production, does not really serve society and its consumers any better than do the family farms".[331] He assumed that the owners of such factories would need more hired workers. What he could not imagine was that the family farm would transform into the "factory in the field", family owned and managed.

He maintained his basic idea about the disadvantages of large farms. In the textbook from 1987, he argued that small farms are at least as efficient as large-scale farms, and tried to prove his point with recounted statistics. Gross sales of products were not a good measure of farm size, as they exaggerate the economic superiority of large farms, and also exaggerate results on firms that purchased inputs as fodder, manure etc. Actually poly-culture on a small firm was often more favorable than statistics showed.[332] He now also attacked the problem from an environmental standpoint. He discussed monoculture, grain deserts, and accelerating soil erosion as a result of agriculture steered by cash cropping.[333]

In his discussion about the small family farm, he stood aside of the main discussion. In the 1980s, American farmers went through a crisis, with a strong tendency to concentration. Dovring was not mentioned even by those who fought for the small farm with similar arguments.[334] Among radical environmentalists today, humans and agriculture are sometimes seen as a disturbance of the natural order,[335] which is an opinion far from the standpoint that was Dovring's.

As a professor in Land Economics at a major American university, Dovring had made some impact especially by continuing his research in agrarian history of the twentieth century, and in pursuing land reform. Still he had wider ambitions.

---

330 Dovring (1969/1970), p. 1 272.

331 Dovring & Dovring (1971), p. 65.

332 Dovring (1987), p. 371–372.

333 Dovring (1979).

334 See for instance Strange (1988), and he is neither mentioned in overviews about the history of American agriculture such as Cochrane (1993) or Gardner (2002).

335 As in *The Fatal Harvest Reader. The Tragedy of Industrial Agriculture* edited by Kimbrell (2002). The authors certainly have a good cause against many drawbacks of modern agriculture, but generally lack understanding of the inherent contradictions in all human activity, as most them are scientists. For humans as a disturbance in the natural order, see for instance p. 66, p. 95.

CHAPTER 7
A Reform program for America (1971–1998)

Theory of waste, 1968

Folke and Karin became citizens of the U.S. in 1968, in the midst of a growing political storm in and outside the nation during the Vietnam War. Folke Dovring even got reactions from other Americans concerning citizenship if not refuting his decision, at least asking questions about it. On April 17th a colleague, after appraisals over Dovring's Mexican investigations (mentioned above), commented upon Folke and Karin's becoming American citizens: "I myself at times feel terribly pessimistic and really ashamed at what is happening in this country." Dovring replied on May 6th: "We have had many negative comments by Americans on America. Much of this I think is based on lack of perspective." He continued that every nation has its strengths and faults. The other professor then concluded the discussion on May 15th with "I guess the less said about your citizenship the better", but he was convinced that the country would be highly enriched by the Dovrings' joining it.[336]

For a time the Dovrings did try to become politically involved in the Democratic Party, but with little success. He and Karin, however, achieved influence against cars parking on the streets in Urbana. But generally politics was not Folke's cup of tea, and he turned to theoretical critique.

Three years after acquiring the citizenship, he and his wife presented a program for a better society, *The Optional Society*, in 1971. He wrote the more theoretical parts, and she gave it a flavor of realism, putting in small anecdotes and examples. I am going to concentrate on Folke's parts. The basic question of the book is: how do we handle an affluent society with a widening range of choice? To enter a supermarket, even in the 1960s, was an overwhelming experience. All these gadgets, all this consumption – why and to what end?

This came to be the start of a project that Dovring probably considered the most important he ever undertook. Had he had the opportunity to comment on my text, he most probably would have suggested an expansion of this chapter. That is the reason why I, when I finished the text, asked a young but erudite economist and economic historian, Gabriel Söderberg, to write an appendix on

336 DA, M.K. to Folke Dovring 17.4.1968, 15.5. and Folke Dovring to M.K., 6.5.1968.

the series of books that Folke Dovring wrote from the late 1960s and during the following three decades (see Appendix 4 and 5). Söderberg's text is a kind of second opinion on this important issue; there will, of course, be some duplication between this chapter and Appendix 4.

The first book, from 1968, sets the agenda for the whole project. The utopian dream of the Dovrings was a society where we all could cultivate our personalities instead of being trapped by the monotonous lifestyle inherent in many aspects of a society based on mass consumption. As all reformers, they were torn between praise of the common man and scorn of his actual behavior. They, for instance, declared that playing bridge could not be seen as a sign of progress. Today such card playing would probably been seen as a good thing, holding the society together. This would be compared with our contemporary society where everyone is "bowling alone" and the social cement is starting to decompose with less social activity.[337] This change of perspective highlights every suggested means to reform the behavior of the people.

A trait he and Karin had in common with other reformers is that they did not say much about the society they wanted to establish, but more about what they disliked in the present society. Two typical American virtues came under attack: the super-rich as heroes to admire and imitate; and the car-society where the automobile is the focal point which determines the doings and dealings of most people.

A fundamental concept in this critique was "social waste", which was waste caused by social factors, both of natural and social resources. In the program from 1971, waste was defined as a result of a society where the opportunities of choice had expanded and affluence, instead of scarcity, became the problem. The problem was especially pronounced in America, which was stamped by the wasteful habits of a frontier society, a society where there always had been plenty of resource to waste.

The concept "social waste" was defined against a goal. If a society is transfixed by cars, as the American society has been for decades, then there is no waste in pouring resources into the car complex (that it causes serious environmental problems is actually a different question). Another, and perhaps less provocative, example could be a medieval society where most people would consider it a good and pious outlay to take from the scarce resources available and build a huge cathedral.

Dovring's goal was for him incontestable: a society should save and preserve resources. This was an ideal of a subsistence economy in which he had been imbibed as a boy. When Dovring presented his ideas around 1970, for many they must have seemed as something from the past. Partly, but only partly, he can be seen as an offshoot from the environmental movement. Dovring pointed out that we live in a time of technological overkill and tend to use technology just because it is available. (A comment is that new technology must be cost-effective, before

337 Compare Putnam (2000).

it is put in use, but Dovring's answer would be that costs are also societal.)

The environmental movement developed early in the U.S., and it had several roots. Many have written the history of the environmental movement, but I am following one of my favorite scholars, Willard Cochrane, who in his overview of the history of agriculture in the U.S. has a short sketch of the growing and changing interest in environmental questions.[338] The movement as a force in society can be followed back to early conservation politics in the late nineteenth century. It started with preservation of natural resources against overuse, but there was also an aesthetical dimension; to preserve areas of scenic beauty, most especially wilderness. An inbuilt contradiction in this goal was that wilderness had to be made accessible to the public, to get the necessary political support. National catastrophes, such as the dust storms of 1933–1935, and the economic depression at the same time, made Americans take more firm action to protect the environment. In the beginning of the affluent society, in the 1950s, the conservationists were pushed aside.[339] However, at the same time, the movement was collecting intellectual momentum, combining new ecological science with the older aesthetical value-system.

A second shock came with the publishing of Rachel Carson's *Silent Spring* 1962, when people realized that something similar to war gas, prohibited during war-time, was being spread in their neighborhoods. Cochrane has a wonderful description of how Carson took scientific evidence from scientific journals "and other obscure places", added indignation and, as a talented writer, exposed it in a best-selling book. As expected, scientist from the hard and agricultural sciences claimed that "she was simply a hysterical woman from the lunatic fringe".[340]

Reading the book is still today a hair-raising experience. The agricultural scientists in Illinois were among those accused. The section about their deeds starts with the following sentence: "Perhaps no community has suffered more for the sake of a beetleless world than Sheldon, in eastern Illinois, and adjacent areas in Iroquois County".[341] Could it be that Rachel Carson was not a favorite among some of the agricultural scientists when Dovring was a fresh professor at the University of Illinois in the early 1960s?

She had collected material for a long time, and, when the storm of critique against her broke out, she stood on firm ground.[342] The public and the politicians demanded action. She lived just a short time after the publication of her book, but long enough to see the victory. It was translated into many languages, and legal actions were taken all over the West.

338 Cochrane (1993), p. 286–295, see also articles in journals as *Environmental History*.

339 Linnér (2003), p. 37–41 describes how the more radical elements in the movement had to give up their positions in the beginning of the Cold War.

340 Cochrane (1993), p. 295.

341 Carson (1970), p. 88.

342 Brooks (1989) about how she worked and the reaction the books caused.

In the 1960s, and even more in the 1970s, environmental issues exploded. However, Dovring, to a large extent, did not participate in the movement, partly because he was not a movement-man, partly because he disagreed. He never quoted Carson. He had read the book, and kept it in his library, but, according to oral information, he was apparently not impressed by it. He thought that she was a bit careless with facts.

His arguments were basically different from those put forward by the preservationists and ecologists. Of course he recognized environmental problems, but mainly he was interested in waste from another perspective: he looked upon the economy as something that must be in balance – where you could not spend more than you have. It was a bit old-fashioned, and had a touch of a household-budget concept. If you wanted to spend, you had to wait until you had saved. In 1978 he declared, talking about the United States' oil imports that: "the U.S. economy behaves like a family that mortgages its house to buy groceries on a standard they could not afford out of current income".[343] And he had a point. Much of social science exists to explain why common sense turned out to be right from the beginning, but, of course, also to control when it is not correct.

In connection with his waste-theory, he preferred what he considered as "a complete economic system", with the British Empire as the model. The U.S. had been such a system until the 1950s, but then became dependent on resources from other parts of the world.[344] He was not criticizing this dependence on the same grounds as modern environmentalists, who have an aversion to the transporting of food and other wares all over the world. Dovring's emphasis was instead more on political and economical suzerainty, and connected with his idea about an economy in balance where scarce resources were not to be exhausted.

Developing the theory, 1973–1981

When the first oil crisis came in 1973, it seemed to support Dovring's conservation-ideology. He started to explore the theory of waste in different ways. Much was published in little known journals, but he started with a success.

Soybeans as a resource had come into his mind previously when he worked with land reform and poor countries. He supervised students working with the possibilities of growing soybeans in northern India. The University of Illinois, and the department of agronomy, takes pride in having helped Illinois farmers become the largest producer of soybeans in the United States. The wide-spread cultivation of soybeans came much before Dovring arrived in Illinois, during the interwar period, and still today corn and soybeans are the preferred crops in the state.

Dovring picked up the soybeans as a solution to different problems. In the

343 Dovring (1978), p. 10.

344 Dovring (1975), p. 2.

OECD Agricultural Review 1973, he argued that imitation meat from soybeans was healthier than meat, and required less land to produce. Soybeans could solve the protein deficiency in poor countries. With rising per capita incomes in such countries, their demand for meat would increase and this could be met with a land-saving production of soybeans as a meat substitute. In 1974, he was invited to write an article in *Scientific American*, where he would be able to further develop his soybean-suggestion. It was presented as the feature of the journal, and Dovring's soybean was on the cover picture. He explained that a growing production of soybeans would make attainable a complete diet without animal products, and thus diminish meat consumption and its associate energy consumption. This was vegetarianism combined with environmental concern. He concluded his article: "If foresight had been applied in the 1950s, for example, it is likely that neither the current energy crisis nor adverse balance of foreign trade would have arisen."[345]

Another theme, which he soon left, was the population crisis. In the early 1970s, Dovring had started to write about the danger of population increase, which he predicted would escalate and become problematic. He, for instance, took part in a meeting in 1975 at Michigan State University, organized by another Swedish scientist, who had immigrated to the U.S., Georg Borgström. As described in a previous chapter Borgström had been pushed out of Sweden much the same way as Dovring, and their speeches had much in common. They, however, never again worked together or even were in contact. Poor countries disappeared from the focus of Dovring's interest when he concentrated his thoughts on the U.S.

In his continuing critiques of the waste-society, he warned in *Illinois Business Review* in 1976 that: "The American economy suffers from a deep imbalance caused by excessive consumption of energy and materials. This is a result of habits formed in the past when resources seemed limitless." He then forecasted that the country would face a shortage of energy within 10–20 years.[346] The "sacred car"[347] was a favorite target in many articles. One ought to drastically cut down commuting by car by "reducing access to parking space",[348] and "urban freeways [was] a mistake".[349]

A reform program that included a closing down of the freeways, and prohibiting parking in the towns was not anything that could win general support in the 1970s. As he now published in small journals, the reactions were few, but angry when they came. In a debate in *Illinois Business Review* in 1981 (February and April), a doctorate student at Dovring's own department, and a private petroleum

345 Dovring (1974), p. 21.

346 Dovring (1976). p. 2, the article was a comment on The Democratic Party Platform.

347 Dovring (1978b).

348 Dovring (1975), p. 15.

349 Dovring (1981b).

consultant answered in separate articles to Dovring's attack on the car-complex. The first mentioned, Paul Peterson, argued that America's oil had not drained out and that the interstate highway system was a defense project to enable the country to move material in times of national emergency. The consult and geologist, Donald Bond, went into more technical discussion; he also declared that domestic oil production had a bright future. Dovring answered in a superior tone, but, as he had been called a "prophet of doom" and "naïve and false", he can be forgiven. He told his antagonists that they used data and logic in an incompetent way, and that Dr Bond had a vested interest as a partisan to the oil industry. "Wishful thinking" was the phrase he this and other times, to brush off his adversaries.[350]

Another thread he picked up and spent much time researching was whether crop production could be used to produce energy: ethanol or methanol. He tried to find the most suitable parcels of land in the state, which would be suitable for the purpose of producing energy-crops.

His whole life, Dovring had questioned simplified measures for development, and now he used this as a wedge to drive into the leading paradigm about energy-waste. He questioned total factor productivity, and, above all, gross national product (G.N.P.). Instead, single factor productivity should be used. His interest in labor productivity and how to measure it goes back to *Land and Labor*, where he tried to get a correct measure for the use and waste of work-time, including aspects such as laziness and incompetence. From the mid 1970s, he started to hold energy productivity as a more essential measure than labor productivity, and this factor of productivity was used as an argument against waste of energy.[351]

He was also struggling with the question about capitalism, which according to Dovring could run a business but not the country.[352] The market is shortsighted, and must be balanced by the leaders of the country.[353] He was gradually approaching a position, which in Europe would have been labeled as being a "Social-Democratic" – and thus, in his mind, he was returning to Sweden.

The warning, 1984

After retirement, a period of intense production followed. It was as if he had spared himself for this thrust forward. He produced half a dozen books that formed a political program about the new society. The first book in the series came in 1984, *Riches to Rags: The Political Economy of Waste*. This book was to be among the most radical he wrote, and it has been told to me that he held it in high esteem. It was, however, not very much reviewed or quoted, to my knowledge.

In the book he found waste everywhere: waste of food, of energy and of hu-

350 See Appendix 1. Bibliography no. 194, 195, also 196 and 205.

351 Dovring (1977); (1979).

352 Dovring (1975).

353 Dovring (1976).

man resources. He launched a theory: [354]

> "Social waste is used consciously, and on a large scale, to sustain prices on things which tend to become less scarce and less expensive. Such waste hurts society as a whole and most of its members, but it favors those who own resources that are wasted. Food, clothing, housing and transportation are all used wastefully so that the owners of land, mines, oil wells, and other basic resources may get a larger share of the nation's income in exchange for what they have to sell."

The owners furthered their interest by promoting more consumption. Importantly he added that the habitual way of thinking as an important factor. The vast resources of the North American continent encouraged wasteful habits that became institutionalized as a part of a multiple locked-in system.[355]

He started with the food system, organized with waste designed to increase consumption. People consumed too much wheat, butter, meat, etc. It was not healthy and if they ate less and wiser – dark bread, vegetable oil, etc. – land could be spared and life more sound.

Then he lanced his most fierce onslaught on the oil-car complex. He saw it as the worst case of social waste, besides spacious houses.[356] He declared that oil was responsible for the foreign policy syndrome, and the way the U.S. was drawn into the dangerous web of Middle East politics.[357] He piled arguments against cars: enormous killing in accidents; waste of time in traffic jams; towns disintegrated; waste of land with highways; health problems because of too little physical exercise, etc. A strong combination of vested interests, encouraging spending and wasteful habits, was the oil industry, the automotive complex and the highway lobby. He again stated the highways as being a mistake.[358] Sadly, he had realized that he was not only fighting the economic interests, but that questioning the car caused heavy resistance on emotional grounds.[359]

Other parts of the book are a catalogue of grievances, held together by a concept of "waste" in a very wide sense. One part is about how talent was wasted by the school system and by discrimination of different groups in the society. Dovring argued that kids spent too much time in classrooms when they could have learned more by private studying and reading. His ideas were a reflex of how he himself was brought up by his father, where he and his brothers were taught at home (see the first chapter). His ideas could have lead to a questioning of mass-education, but he did not elaborate his thoughts in that direction. He was against

354 Dovring (1984), p. 1.

355 Dovring (1984), p. 3, 6.

356 Dovring (1984), p. 26.

357 Dovring (1984), p. 35.

358 Dovring (1984), p. 59–60.

359 Dovring (1984), p. 53–54.

a costly defense and weapon industry, but accepted that large powers had to react against "small gangsters on the international scene".[360]

In discussion about drugs he proved that he now was more of an American liberal, than a Swedish one. Prohibition was not the solution to drug problems, according to Dovring. Drug tolerance in Sweden was (and is) zero, and there is nearly no discord about this. Had he been still in his old country, his position would have been nearly as odd as the position he took about cars in the U.S. More and more his general standpoint on society, however, took a Swedish outlook. He even talked about the need for a "guided market economy".

A paradox with dark prophesies is that the more wide spread the message is, the less likely it will be correct. But the reverse is not always the case. A doomsday prophet does not need to be right because no one is listening. Hardly anyone listened to Dovring, but was he right? I am going leave his grumbling over the school-kids, laboratory animals, etc. aside, and concentrate on waste of food and oil.

In mass-consumption combined with capitalism there is no incentive to produce less, no one will gain or make a fortune by selling less. This has given us welfare but there is also a downside. Over-consumption of food is disabling and even killing a large part of the American people, and it has become a lifestyle. Companies in the food chain have no incentive to stop this ongoing degrading of people's health and life. Dovring pointed out this problem, even if he could not foresee how serious the illness has become today.

Even more serious is the problem with energy, which causes environmental change and violent conflicts. Oil and gas will run short in a not too distant future, and is no longer as cheap as we have been used to. Not only the United States, but also the whole world, will eventually be forced to give up – at least parts of – the oil-car complex. Different alternatives present themselves for us. One is that new kinds of fuel will allow us to keep the cars, the highways, etc almost as today. Another possibility is the establishment of a totally new system for land transport, and this will need and cause a different organization of the society. A third alternative is, as Dovring anticipated, catastrophe and crisis. The less we think about the problem, the more probable the third alternative will be.

The literature about the oil-car complex is enormous and critique of its various aspects is abundant. Many defend it, not the least economists. Dovring is never mentioned in the discussion of today, but the total onslaught on the oil-car complex he launched has few parallels.

Textbook and controversy, 1984–1987

In several editions he had published his syllabus for the class in land economics as a mimeographed book, and, as such, it was quoted in the scientific literature.

360 Dovring (1984), p. 126.

When he retired he tried to get a publisher. An interesting correspondence between Dovring and his publishers started where we get an idea about critique from traditional economists.

Dovring had sent in a large manuscript to the publisher Jay Bartlett, at the Breton publishers, and he had passed it to two reviewers. In a letter in May 1984, he asked Dovring to make amendments that the reviewers had suggested (their reviews are not preserved in the Dovring archive and we do not know their names). Bartlett wanted to make the syllabus into a textbook for "the marketplace", and his letter regarding the matter was rather kind and encouraging, even if demanding.

Both reviewers had asked for more basic economic principles and one of them asked for more applied microeconomic theory. As Bartlett remarked, this person "seems strongly inclined to forgive the differences [in position between him and Dovring] if you'll just give him some microeconomic applications to play with".[361] Bartlett also asked for "more glue" in the text, with stronger structure and better theoretical underpinnings. A further remark was that Dovring blurred the distinction between normative and positive statements. Bartlett did not ask Dovring to skip his normative observations, but to label his opinions more clearly and to include alternative opinions. The reviewers, and Bartlett also, wanted a series of other amendments: as more examples, more documentation, more additional graphs, suggested readings and discussion questions for the students. Finishing his letter Bartlett indicated that this could be seen as a starting point for negotiations.

Folke Dovring could not take this criticism lightly. As an answer to Bartlett's rather informal letter, Dovring, a week later, wrote a strict and angry letter. In Dovring's opinion "micro theory used in isolation is a polite term for consumer fraud". He argued that the market for the book was not adherents to Reagan-economics (an experiment that, by the way, would collapse in four years), but institutional economists. Dovring also takes a deep breath, and declared that he was not prepared to submit to monitoring, after "the many books and the load of other writing (in this country and across the world) that I have below my belt".[362] Furthermore, adding practical pointers was alien to his style of teaching, where he concentrated on the general, durable knowledge. At the end of the letter he became more cordial and he liked what one of the reviewers had suggested, something he actually already planned to include in an extended version.

Bartlett seems to have dropped the whole thing and another publisher at Breton took over, Edward Francis. He sent out an expanded version of the text to two reviewers, one a "respected professor at a major Midwestern university" and the other from "a top university".[363] In his letters commenting on these reviews (nei-

361 DA, Jay P. Bartlett to Folke Dovring 1.4.1984.

362 DA, Folke Dovring to J.P. Bartlett 7.4.1984.

363 DA, Edward L. Francis to Folke Dovring 24.9.1984 and 21.10.1984, in the last letter he also mentioned that the second reviewer "has more favorable things to say".

ther of them preserved) in September and October of 1985, Francis reminded about the market: "Our goal is to produce a textbook which is competitive with the leading best sellers like Barlowe in undergraduate courses".[364] Raleigh Barlowe's textbook came in three editions between 1958 and 1978, and must be considered as a best-seller. The reviewers called for more basic principles (micro economic concepts), and the publisher commented that Dovring ought to "give faculty what they believe is best for their class".

Again Dovring was not happy, he rejected the comments "on professional grounds", and would neither write, nor use in class, a book of the kind that the reviewers had suggested.[365] There had also been a threat in the letters to Dovring. The publisher asked Dovring to find another publisher,[366] but Dovring stood his ground, saying his contract was binding and answered that he had talked with his lawyer.

The preconditions for a bestseller were apparently not there. The publisher just wanted to print the book and bury the project. Some years later, in 1989, Dovring got a letter from the editor of another company, Delmar publisher's inc., to which the book was transferred. It had been rumored that the whole edition was lost in this transfer, but that was not true. However the book "is for a very specialized market".[367] Dovring was told that only 54 copies were sold this year, two years after it was published. Today the book is found in few libraries, and even the single copy at the library of the University of Illinois seldom leaves the shelf.

The book has seldom been quoted, but it got one rather favorable review by a prominent agricultural economist in the well-known *American Journal of Agricultural Economics* 1988. Gene Wunderlich, at the USDA (the U.S. Department of Agriculture), wrote in *American Journal of Agricultural Economics* 1988 that the different subjects were clearly stated, with simplicity and clarity, and he praised Dovring's imaginative thinking and scholarship.[368] According to Wunderlich it was a personal text, yet free from bias. Then he added: "His private war with the automobile, for example, is not discernible". This is an interesting comment, as it implies that Dovring's questioning of the oil-car complex had been turned into a joke among those economists who knew of him.

The book is about 500 pages and probably gives a fairly good idea about his lectures, which were quite popular among some of the more intellectual students.

364 DA, Edward L. Francis to Folke Dovring 24.9.1984.

365 DA, Folke Dovring to Edward L. Rancis 4.11.1984.

366 DA, Edward L. Francis to Folke Dovring 21.10.1984, Francis told Dovring that "we are no longer publishing in Agriculture" and he concludes the letter: "I again urge you to consider placing your manuscript with a publisher which is active in the field".

367 DA, Mar W. Huth to Folke Dovring 25.10.1984.

368 Dovring had published in USDA publications edited by Wunderlich, for instance Appendix 1. Bibliography, no. 154, 155 from 1976 and 212 from 1983.

As one could expect after the discussion referred to above, he started, with an attack on micro economic theory, and he stated that the economy, as a whole, cannot be understood as the sum of its many small parts. Engel's law and the law of diminishing returns were presented as fundamentals. He also presented Malthus, Ricardo (a favorite), von Thünen and Marx, but not neoclassical theory. He actually was against neoclassical theory and declared himself as an institutional economist.[369] Environmental questions have a long section, where he for instance mentioned Betty Meggers and her work on the Amazons.

Plenty of text was spent on small-scale farms' being as efficient as large-scale farms. Still he knew that he was fighting a losing battle, and blamed those "who pay more attention to machines than to people".[370] Historical facts and observations were strewn in, such as that survey methods used in the North America in the nineteenth century can be traced back to techniques used by Roman surveyors, proven with maps indicating similarities in the layout of land. His credo about waste of resources was mentioned throughout the book. If the U.S. took on a less resource-intensive way of life: "our society can instead continue its original mission in the world as the society of equality under freedom."[371]

The year before Dovring's textbook came out, the fourth edition of Raleigh Barlowe's bestseller *Land Resource Economics* was published, in 1986. This totally destroyed any possible market that could have existed for Dovring's book. Barlowe had everything reviewers had asked for: much micro economics (with many diagrams with crossing curves, so typical for economists); statistics; examples; maps. Barlowe presented hundreds of articles and research results in his book, and it is useful as a handbook and book of references. He never once refereed to Dovring, not even when it could have been appropriate, as in the text on land reform.[372]

Barlowe wrote a balanced textbook, trying to give every opinion their fair share, but he held opinions alien to Dovring's. For instance, Barlowe argued that the United States has skimmed the cream of its stock of natural resources, but there had been nothing unnatural in this. Fossil fuels power our present industrial society, but he admitted that, in the future, the shift must be made to some other solution.[373] Barlowe mentioned the many "gloomer-doomers" coming forward around 1970, but he also wrote that "wasteful uses of resources" should be avoided.[374] This is far from Dovring's theory of an intrinsic tendency to waste in our society.

369 Dovring (1987), p. 321.
370 Dovring (1987), p. 384.
371 Dovring (1987), p. 194.
372 Barlowe (1986), p. 504.
373 Barlowe (1986), p. 90–91.
374 Barlowe (1986), p. 18.

The reform program presented, 1987–1988

The textbook was actually a sidetrack, as his main task was the reform program. Dovring was busy pressing forward with the series of books he had planned. All of them had, as the book from 1984, the subtitle: *The political economy of …* Certainly he was referring to forerunners as John Stuart Mill and Karl Marx, and he probably considered his project as being, at least close to, the dimensions of these scholars.

In 1987, he presented an attack on the prevalent methods of measuring production, and two books about agriculture came in 1988 – one of them about production of fuel and the other an overview of world agriculture. In 1991 he argued for a more equal society, and in 1996 he further defined his position while criticizing communism.

In the 1987 book, *Productivity and Value. The Political Economy of Measuring Progress,* he came back to one of his favorite ideas, that total factor productivity as a way of measuring productivity tended to obscure system boundaries and cause problems with aggregation. Specifically he attacked the G.N.P. In addition he also revealed a new argument: "the analysis serves to justify the status quo rather than to explore how it might be changed."[375] Furthermore G.N.P. did not always measure the important things, and "the spontaneous tendency toward social waste out of affluence becomes ingrained" in this measure. He asked: "how important are the things which conventional productivity research measures?"[376] Here a social reformer, and not only an economist, spoke.

The book on fuel and agriculture, *Farming for Fuel. The Political Economy of Energy Sources in the United States* in 1988, gave suggestions of how to solve four problems in one stroke: petroleum imports, farm surplus, soil conservation and air pollution. Oil was running short, other energy sources (as nuclear power) had too serious drawbacks, and thus biomass was the only sustainable source. He went into detailed technical matters; methanol (in his earlier writings ethanol), costs, acreages, etc. He demanded change: "The national administration in 1987 emphasizes market forces rather than planned public policy intervention."[377] The last lines of the book read: "There is certainly no time to lose."[378]

In the second book on agriculture, *Progress for Food or Food for Progress? The Political Economy of Agricultural Growth and Development* in 1988, he was more descriptive, and presented it as an overview for the general reader and a basic text for graduate students. He thus joined a tradition of writing world overviews of agriculture, with other famous books by, for instance René Dumont (1957) and D.B. Grigg (1974). Dovring tried to return to methods he developed early in his

375 Dovring (1987), p. 154.

376 Dovring (1987), p. 166–167.

377 Dovring (1988), p. 127.

378 Dovring (1988), p. 132.

life, collecting huge amounts of facts and presenting them in a systematic way. However, his overview is not as exhaustive and factual as those he made in his younger years. He did not try to construct new measures for certain traits, and stuck to a more narrative form of presenting his material.

As in many other books, Dovring inserts small pieces of historical information, but the most interesting part of the book is a number of regional overviews, where he was pessimistic about most of the world.

China attained early political stability and therefore, built up an early high population density. *India* had political stability later and thus experienced a later demographic upsurge. His main conclusion was that most parts of Eastern and Southern Asia were far from an industrial and economic breakthrough. *North Africa and West Asia* suffered from cultural and political traditions, and the desert character of the environment was also limiting. *The rest of Africa* had an especially unfavorable natural environment. The soils were old and leached, humus was destroyed by the tropical weather, erosion was severe and diseases were endemic and debilitating. The social structure was determined by tribalism, which was a hindrance. *Latin America* had institutional problems in the extreme concentration of landholding, which had left vast areas of farmland underutilized, and also caused an unnecessary urbanization. *Eastern Europe* was blocked by the command economy, and he predicted that agricultural problems called into question whether the USSR could continue as an autarky. *Western and Southern Europe* utilized technologically advanced family-scale farming in many regions. The farm structure was fairly stable, even though the countryside had been depopulated in the 1950s. In *Northern America,* the plains were the largest, coherent temperate-zone agricultural region in the world, thus was the most important food producing area in the world. The export of these products had mainly been positive to the rest of the world. Overproduction was the main problem, and the solution was diversification of land use (producing fuel etc).

He started and ended the book with the acceleration idea he had presented in the 1960s – preindustrial population increase was accelerating, and by inference, technical progress must also have been accelerating. Technical progress had developed at a quickening pace and: "all peoples have been moving toward the industrial revolution, though at somewhat different paces and with various reversals."[379] Today this change has given us more choices, and we do not need to optimize the use of the world's resources.

Some of the predictions proved to be correct, such as the falling down of the Soviet empire, others not, such as Southern and Eastern Asia being far from an industrial breakthrough.

His next book, *Inequality. The Political Economy of Income Distribution* in 1991, presented another part of the reform program: a society that ought to have more concern for the poor. He introduced new ways to measure economic distri-

379 Dovring (1988b), p. 7.

bution, with use of the exponential function. The main argument against inequality was that it destabilizes society. Dovring enumerated a lot of assumed consequences: deficit finances, environmental problems, and a degrading knowledge among the young.

He saw a nation with waste of oil, waste of environment, waste of food and growing inequality, which he considered as social waste. American conservatism was nothing but wishful thinking. He asked the question: productivity for what? His pessimistic answer was: "Waste is good for profit because it increases scarcity".[380]

Dovring's theory of waste was in many ways an old-styled conservatism and his ideological roots go back to early twentieth century agrarianism, in Europe and in the U.S. He had no intellectual link to the environmentalist movement, even though his standpoint agreed with theirs on several points: equality; small-scale production; environmental care; critique of the oil-car complex. For instance, in October of 1993, he gave a speech on "Alternative fuels: Alternative to war" in Mumford Hall at the University of Illinois, the very building where he had spent so many years as a professor.

His last book in the series, *Leninism. Political Economy as Pseudoscience,* published in 1996, was an assault on communism in the Leninist-version. He had always been a fierce fighter against communism and the Soviet Union; early on he realized that the social system was dysfunctional, with a social waste much larger than in the West. According to Dovring, Lenin was never a scientist, just a politician in everything he wrote. What he produced was pseudoscience, which became disastrous for Russia.

Dovring is much more positive regarding Marx, as a supreme thinker on political economy. He concludes the book with the statement that "some elements of Marxism still stand", especially Marx's theory of rent, and also "his warning that too sharp class differences are a deadly threat to all of society".[381]

Reviews of the reform program

His first book in the series, from 1984, was not reviewed, to my knowledge, but the others were in a dozen different reviews in journals registered in ISI, the citation index. (I could not utilize Dovring's own collection of reviews as he had none of them on the later books in his collection.) They give a hint of how his books were received, and why they fell out of most of the ongoing discussion.

His book about measuring production from 1987 was reviewed in *Economic Development and Cultural Change* by John Kendrick, who admitted that aggregated economic measures had limits, but that they must be used. Much less sympathetic

380 Dovring (1991), p. 131.

381 Dovring (1996), p. 146, but he also states that "Of Leninism, by contrast, nothing of positive value remains".

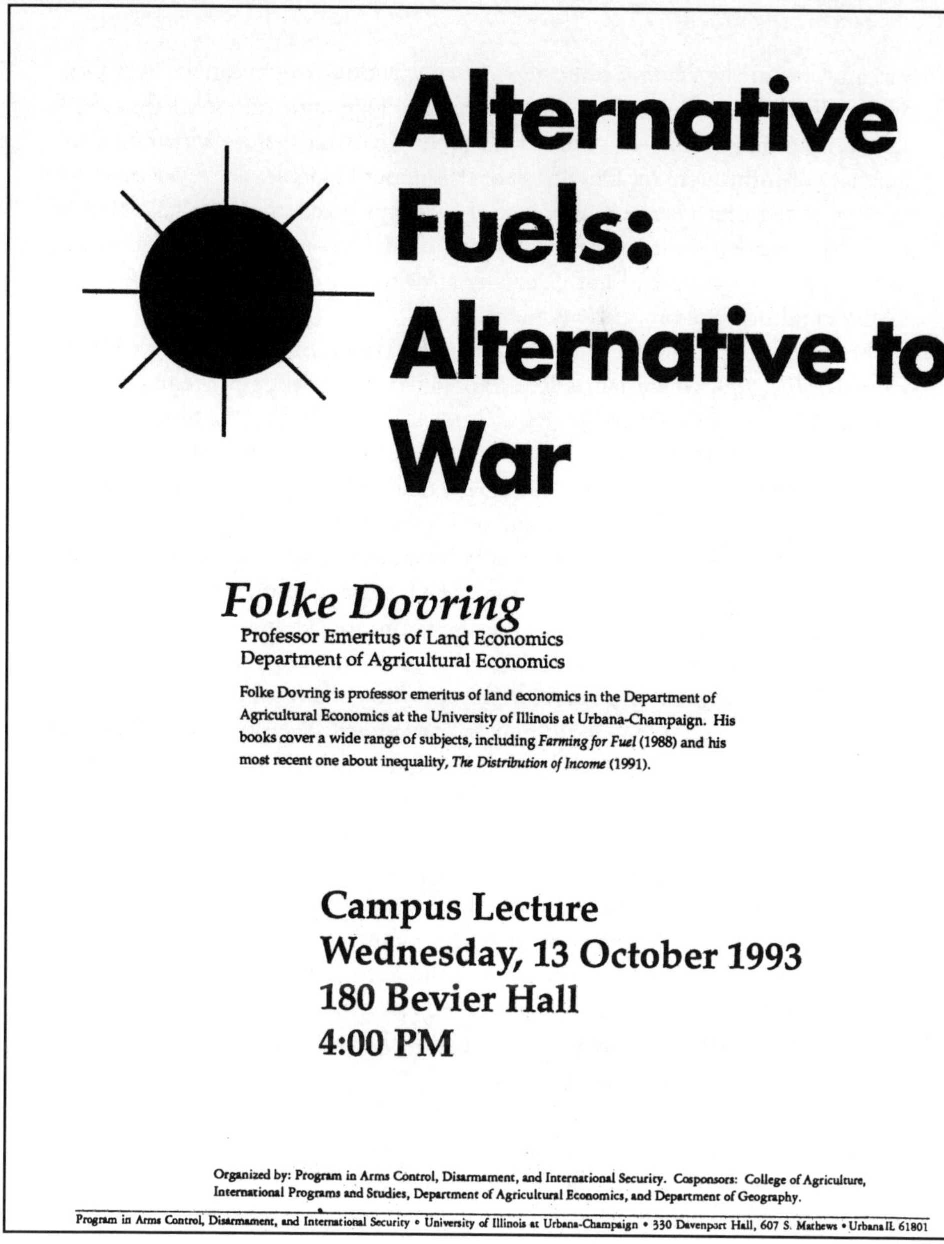

Alternative Fuels: Alternative to War

Folke Dovring
Professor Emeritus of Land Economics
Department of Agricultural Economics

Folke Dovring is professor emeritus of land economics in the Department of Agricultural Economics at the University of Illinois at Urbana-Champaign. His books cover a wide range of subjects, including *Farming for Fuel* (1988) and his most recent one about inequality, *The Distribution of Income* (1991).

Campus Lecture
Wednesday, 13 October 1993
180 Bevier Hall
4:00 PM

Organized by: Program in Arms Control, Disarmament, and International Security. Cosponsors: College of Agriculture, International Programs and Studies, Department of Agricultural Economics, and Department of Geography.

Program in Arms Control, Disarmament, and International Security • University of Illinois at Urbana-Champaign • 330 Davenport Hall, 607 S. Mathews • Urbana IL 61801

Flyer for talk of Folke Dovring on fuels and war 1993.

was a review in *Agricultural Economics Research*, a USDA-publication, with Gene Wunderlich as editor (who was rather favorable to Dovring's textbook). The USDA was at this time working with multifactor productivity indexes for agriculture, and their lack of enthusiasm for Dovring's attack is understandable. The reviewer, James Hauver, stated that Dovring's proposal of a disaggregated approach fails, that he was mistaken about indexes, treats labor and capital inconsistently, etc. According to the review, Dovring had not given any attractive or credible alternative to the USDA's multifactoral productivity measure.

The suggestion to produce fuel on farmland was enthusiastically reviewed in the journal *ISIS* 1989, where James Rees thought that Dovring's program "deserves careful attention". His only critique was, ironically enough, that the historian Dovring lacked a historical perspective. The reviewer in the more official *American Journal of Agricultural Economics* 1990 agreed that a reliance on oil from the Persian Gulf carried an overhead in military and political costs, but Fred Hitzhusen thought that the author's strong advocacy resulted in subjective statements. The reviewer referred to his own, and by Dovring overlooked, research in the field.

Dovring's second book on agriculture was surprisingly attacked from the left, because it was not factual enough. In *American Journal of Agricultural Economics* 1989, Richard Norgaard and Lori Ann Thrupp wrote that the book had no theory, not even a theme. As Dovring did not accept cultural diversity, his thinking was outdated in relation to third world peoples, and, for instance, his explanation of Africa's backwardness with tribalism was one of several "simply indelicate assertions". In *Geographical Review* 1989 Daniel Weiner wrote that the strength of the book is its historical breadth, but Dovring shows little appreciation of indigenous agricultural knowledge. Weiner's conclusion must have astonished Dovring: "the book outlines neoclassical economic arguments on global agricultural development."

The book on income distribution and the need for more equity from 1991 got, as suspected, critics from the right. Greg Duncan declared, in *Journal of Economic Literature* 1993 that he was frustrated by the lack of detail in both fact and analysis, and that the arguments were on shaky ground. The reviewer was not pleased by recommendations about more control over the economy. Even more slashing was Dominick Salvatore in *Southern Economic Journal* 1992–1993 who was "not impressed with this book". A third review, but only short and neutral, was published in *Growth and Change*.

His last book in the series, about Leninism, got the worst criticism. Tom Erwing in *The Russian Review* 1997 wrote that readers seeking an understanding of Lenin's influence "will want to look elsewhere for their insights". Jules Townshend in *Europe-Asia Studies* held it as "a far from serious book"; several important writers had been left out of the discussion.

Perhaps the conclusion could be that Dovring was in the midstream, as he got critique both from the right and from the left. But the truth is rather that he was off-stream, holding ideas that did not fit in anywhere.

Philosophy and history, 1998

His last book, *Knowledge and Ignorance. Essays on Lights and Shadows,* is of a different character. He leaves his writings in the field of social and economic reform, and becomes more philosophical with questions about epistemology, and more precisely the question: what can we know? (Further on these questions see Appendix 5 by Gabriel Söderberg.)

He starts with a memory: "Growing up in the family of a famous poet in the middle of old-fashioned peasant country (Sweden), I experienced nature more intensely than do most modern people."[382] His interest in agrarian questions and environmental concern goes back to his feelings when he, as a lonesome boy, was strolling in the nature and inhaling freshness and freedom.

He appears humble, and declares that: "Respect for the unknown should make us to hold back judgment more often than we do."[383] He then touches on a number of subjects. Some of them are familiar, for instance his pledge for the peasants and their life style and wisdom. This was the ideological basis for his life-long struggle for the small farmer. He also takes up a spectrum of other questions such as animal feelings and anthropocentrism, occultism and parapsychology, different forms of energy and the entropy law. He is suspicious of general natural laws and also of Darwinism whose mechanistic approach overlooks "the toolbox of creation".[384] However, he does not end up with a God of any traditional religion, but rather with questions about an intelligent cosmos.

Towards the end, the book changes character, and he makes a plea for the need of history, opening with a favorite quotation from the French historian Marc Bloch: "There is only one science on mankind through time" – namely history.[385] He returned to where he began. The hard core of historical sciences is the study of archives, and Dovring stresses the need for source critics. He also states that one should never oversimplify history, there is always the unexpected. And one last time he confronts his old adversary, Erik Lönnroth:[386]

> "the intellectual mistake that is common among humanists, particularly in Europe: the belief that intuitive insight can ever become a category of scientific knowledge without ex post facto testing of its results by objective criteria. There are numerous examples already among historians, who all too often accept an intuitive hypothesis merely because it was formulated by an acclaimed historian who is generally believed to have strong intuitive powers."

382 Dovring (1998), p. 1.

383 Dovring (1998), p. 3.

384 Dovring (1998), p. 107.

385 Dovring (1998), p. 15, "Il n'y a qu'une science de l'homme dans le temps".

386 Dovring (1998), p. 134.

Half a century had passed since the big battle about agricultural history in Sweden, but for Dovring the case was not closed. And why should it be closed? This was the most important battle of his life, which forced him into a new career and into exile – an exile where he again took a position removed from the mainstream.

CHAPTER 8

Living has an end

How can one assess a life? Most biographies are about the mighty and important, about Napoléon and Roosevelt, and this is understandable because they have had more influence over history than others. Yet a functioning society consists of junior schoolteachers, nurses, shop assistants and many others, and every person's life could be described as a novel. One must have the right to write the biography of every human; the life of an ordinary person will reveal something about the society as a whole. We are all pieces who mirror big history.

Dovring is something in between, not an ordinary man, but neither very famous. He can be treated in his own right and also as a part of trends in history. I have tried both, and my concluding chapter starts with an estimation of his influence as an entrance to some summarizing remarks. One observation is that he was comparatively less successful when he worked in America with American problems, than he had been when he researched European agricultural history or modern agricultural history of poor countries, as Mexico.

Three large processes

Three major societal processes can be studied through the biography of Dovring. One involves European conservatism in the academic world. I have already extensively written about this in earlier chapters. During his first period as a scholar, his focus on the history of the common people and quantitative history on the one hand lead to his rejection, but, on the other hand, helped to establish a new orientation for Swedish historians.

Much of the inspiration for the reorientation came from the U.S. The second important process Dovring took part in was American expansion and take-over in the intellectual world system. From the 1950s, the United States played a totally dominating role, not only in popular culture, but also among the universities in the world. My assumption is that this is the real basis for the leading role that the United States has had for decades in the world society. Economic and military power has had certain importance, but they are not the only, or even the most important, means of controlling the global society. America's share of the world production is constantly falling, and, since World War II military success has been limited. To put it more concrete: the United States could not win the war in Vietnam, but could produce Hollywood movies about it, and thereby influence our idea about how the war was fought.

To understand how the United States acquired this position, at the very moment when a cultural and intellectual global world community was formed, the Dovring case is a good example. The University of Illinois' snatching him after *Land and Labor* and his many articles in the field of modern agricultural history, is a part of this success story. We can find many examples of such forceful activity in the American university system from the 1950s to the 1980s, when intellectual freedom and academic competitiveness are added to the picture.

What was the third large process? According to my opinion, his not very successful project during the last decades of his life, when he proposed a reform program for the United States, could be a sign of a blind spot in the structure of the U.S.

Isolated in the United States

Many of the indications of influence I have used (see Appendix 3) are imprecise, but, taken altogether, it appears that his publications, up to the mid 1960s, had more influence than his later publications. Was this because of his shortcomings, or was it something in his program that became a barrier?

In an earlier chapter, I discussed his failure in the Swedish academic society, and I concluded that this only partly could be blamed on Dovring's lack of social competence. A university needs diversity; it needs people who promote unusual standpoints. They can be inspirers for others, who later take up the issue. In America, Dovring early on drew attention to issues that were subsequently taken up by other scholars, such as energy production by grain. Nevertheless scholars today working with this issue have certainly never heard of Dovring.

An indication of his isolation is that he often did not relate himself to the other leading scholars in the field. This tendency can be found already in his early writings, but later in life this habit became highly evident. Writing about environmental questions, he did not quote or even mention Rachel Carson.[387] Writing about India around 1970, he did not mention Gunnar Myrdal.[388] There are many other examples, such as talking about von Thünen without mentioning

387 In his chapter about "Conservation of environment" in *Land Economics* (1987b), p. 449.

388 Myrdal had finished his important study about Southern Asia in 1968. Later, in Dovring's book about world-wide agriculture from 1988, Dovring has Myrdal in the reference list, which probably refers to when Dovring, with dislike, talks about "foreign commentators" who believe that tractors would be of use in India, see Dovring (1988), p. 229, compare Myrdal (1968), p. (1277–1277). His relation with Gunnar Myrdal was somewhat complicated, he did not mention Myrdal when he wrote about race relations in the U.S. and in a review he praises the Swedish economist Assar Lindbeck as representing the "post-Myrdal" era. On the other hand, Myrdal in his long discussion on underemployment in the countryside in *Asian Drama* (1968) did not mention Dovring.

Michael Chisholm, and discussing Malthus without mentioning Ester Boserup.

The explanation for omitting all these references could be that they were not important for his specific questions. But most probably he did not quote people he disliked. This strategy is rather common in the academia, but it is a dangerous strategy as it arouses incredulity among the peers. This was also a point some of his later reviewers made.

It is quite natural, considering the overwhelming amount of books and publications produced every year, that only a few can be noticed and remembered. Still his series of books were published with a well-known publisher, and they had qualities in reasoning, as many of the reviewers also admitted. Perhaps his lack of real success in later years simply was the result of earlier publications' being better. He did not present the overwhelming amount of data which was his trademark when he had studied medieval or modern agricultural history.

To point at his position as an immigrant as an explanation is not sufficient. The United States of America is perhaps the country in the world where an intellectual immigrant has the best chances to succeed. There are no harsh language-barriers, as many Americans have English as a second language. The university system, and much of industry, depends on a flow of immigrants, and thus America has built up a social structure adapted to an influx of millions of new inhabitants. There is a feeling of being accepted to which most people coming to the country can attest. All this does not, however, nullify the problems that every immigrant has with adaption. He or she has to learn to think in a new way and accept new habits. Dovring was not prepared to accept the American mentality, and especially not a habit he identified as being a waste of resources. This then brings us from his position as an immigrant to his ideological position.

One can for instance compare with Dovring's counterpart, the Swedish scholar Georg Borgström, who was forced to leave his scientific position in Sweden in 1955–1956 and emigrated to United States the year after. The reason why he had to leave Sweden due to his ideas of conservation, very much inspired by American scholars as Vogt and Osborne. As told earlier, Borgström had a position as the director of a Swedish research institute partly financed by the package industry. The executives of the package industry acted to get him dismissed. Indeed, his loss of the office was followed by an intense debate in Sweden. He was then recruited as a professor at Michigan State University 1956, and became a citizen of the U.S. in 1962.

Borgström was successful in spreading his message. He published many articles and several books about the population-resource problem, and about the threatening over-population in the world. He drew attention to a lot of questions, not the least of which regarded over-fishing in the oceans. His books were published in many languages and in several editions. In the late 1960s and early 1970s, he was one of the most influential debaters in the world. He was a part of a whole wave of discussion about environmental questions, preservation, the world as a household, etc.[389]

Why couldn't Dovring ride on this wave? Could it be that his ideas were too unrealistic? His attack on a society structured around the car (free-ways were a mistake), tended to turn his reform program into a joke (the agricultural economist Wunderlich talked about Dovring's "private war" with the car). Perhaps Dovring was not taken seriously, and thus his theory about the waste-society also fell into oblivion?

He was certainly not the only one demanding radical change, but generally the environmentalist debate during the last fifty years has tended to develop mainly within the system boundaries. When conservationists after the war started to talk about the problem with the free market, they ran into serious ideological problems. In the early 1950s, the leaders among instead turned to overpopulation as the most important problem.[390] (It is interesting that this change of the conservationist debate occurred simultaneously as land reform was abandoned as the goal for American international activity.) The overpopulation crisis has today been more or less solved; people the world over have better nutritional standards today than in the middle of the twentieth century. The other large environmental issue, which Rachel Carson took up, about poisoning of nature, is also on its way to be solved as new technology has been developed. Humanity poses itself problems it can solve.

Perhaps Dovring's failure partly occurred because he took up questions that were nearly unspeakable. His critique of the oil-car-complex would have demanded a totally different logistic structure. Today we are facing serious problems of various kinds: war; world-wide environmental problems such as climatic change; resource depletion. Many of these related to the car-oil complex, and to what Dovring labeled a "waste-society". Maybe Dovring in the future will be looked upon as one of the great foresight, if he is remembered at all.

Leaving the cause of his failure as an unsolved question, I, at the same time must admit that he, as a professor, was allowed to research and publish far out of the mainstream – which is a proof of the strength of the university system.

Living his theories together with Karin

When I talked with people in Champaign-Urbana regarding what they remember about Folke Dovring, and his wife Karin, they always started with same sentence. I have heard this so many times that it must be like a reflex, or an attribution that

389 About Borgström, and the international discussion he was a part of, see Linnér (1998) and with further international comparison in Linnér (2003). See also above on examples of scholars leaving Sweden in the decade around 1950, where Borgström is one of the examples.

390 Linnér (1998), p. 123.

neighbors, colleagues and others make when they hear of this couple. They tell me: "They did not have a car."

This sentence usually develops into a description of the old couple walking together with a shopping bag, or a statement that they always were dressed up, even in the worst summer heat. Yet it is always the non-car-sentence, which comes first. This is extremely astonishing, and interesting. It explains a lot. About America.

For most Europeans "not having a car" would not be considered as the most fundamental trait in a personality that people would mention when they speak about that person. Still the European society is structured around the car, with streets, roads, gas stations, energy flow, etc. The different view does imply that Europeans are not as obsessed with cars as are most Americans, many of whom does not touch a bike or enter a bus after their sixteenth birthday. I am well aware that this is not valid for very poor people or for some of the great cities on the east coast, Chicago and San Francisco and a few other dense cities.

When I came to stay in Champaign-Urbana for a while, writing this book, I wanted to test if it were possible to go everywhere on a bike. And it was. I could ride even to the mall outside the town, even though sometimes I had to ride on the sidewalk. The real problem with my experiment was not the possibility, it was that I was considered a little bit strange. I was simply not adapting to the culture.

The Dovrings not only abstained from a car (with the empty garage as the obvious sign for the neighbors), they ate mainly vegetarian, with soya beef, and they invested, not in the stock market, but in land. In his garden, Folke Dovring tried to grow a Little Sweden, with plants similar to those he remembered from home. Every lunch he walked home for a piece of food and a chat with Karin. It took him about ten minutes to go back and forth to the office, it was a pleasant walk in a small wood with squirrels jumping all around.

Karin Dovring worked at home, initially for Yale with research and investigations about propaganda. Later on she started to write short stories and poetry in English. She became quite successful in her new occupation, and she is absolutely worth a biography of her own. In some of her short stories, she comments upon their life. There is a story of an academic who tries to become a politician, and, after his failure, his wife tells him that he should stick to what he does best. This is a remark about her husband. She looked upon American society with humor. And she made a poem that Folke Dovring loved so much because he thought it described his life:[391]

I know only this/ The greatest bliss/ Is not to receive/ Or try to retrieve/
But nevertheless/ Make life possess/
A glimmer of light/ Despite /All darkness and fight

391 Dovring, K. (1995).

His life had glimmers of joy in the darkness, and I would assume that he was very much connected with a mini-Swedish life – on his own premises – he lived with Karin. He was happy with the freedom of thought and writing he had, and with all the time he spent in the wonderful university library in Urbana, learning new things all the time.

Karin also told me, during the work with this book, that Folke Dovring talked about himself as "The monk from Skara", which alludes to another poem, which is quite famous in Sweden. The author who composed it was Gustaf Fröding, and the poem is about a run-away monk who had happened to slay one of the canons (Lasse Canonicus). The monk was severely punished, and when he got away he was filled with hatred. But then, in the poem, he changes his mind and declares his love of all living beings: "nobody is evil, and nobody is good." Everyone should try help everyone else, and give a hand to brothers who fight in the flood of evil.

The runaway scholar, who had slain one of the leading professors and was punished and forced away, tried his whole life to save humanity and stand up for good values. That was how Dovring liked to look at himself.

APPENDIX 1
Bibliography

Folke Dovring, Published writings

Summaries, etc. are given in brackets.
Commentars about second editions, etc. given in brackets after the entry.
Mimeo = mimeographed.

1945

1. Fogdö (Vårfruberga) klosters jordebok. En obeaktad källa från tidig medeltid, med kommentarer. In: *Vetenskaps-Societeten i Lund, Årsbok* 1945, p. 81–104

1946

2. Boskaps- och utsädeslängdernas uppgifter rörande utsäde. Ett bidrag till en kritisk värdesättning. (Summary: The seed schedules of the seventeenth century.) In: *Svensk Geografisk Årsbok* 1946, p. 39–50

1947

3. *Attungen och marklandet. Studier över agrara förhållanden i medeltidens Sverige.* (Summary: Attung and markland, studies on agrarian conditions in Medieval Sweden.) Ph.D. dissertation. Lund 1947, 371 p.
4. Attung. In: *Svensk uppslagsbok,* second ed., Malmö 1947, vol. 2, col. 764
5. Fogdö kloster och dess äldsta jordebok. In: *Sörmlandsbygden* 1947, p. 85–90
6. "Gilzla oc grutha" i Smålandslagen. In: *Arkiv för nordisk filologi* 1947, p. 258–260

1949

7. Järnskatten i Gästrikland. In: *Blad för bergshandteringens vänner* 1947/48, p. 153–158
8. Une partie de l'héritage littéraire de Grotius retrouvée en Suède. In: *Mededelingen der Koninklijke Nederlandsche Akademie van wetenschappen, Afd. letterkunde* 1949 (N.R. 12:3), p. 237–250
9. Review of: Björn Helmfrid, Tiondelängderna som källa till ett byalags ekonomiska historia 1555–1573 (Stockholm 1949). In: *Östgöta Correspondenten*, 23 April 1949

1950

10. Anteckningar om drottning Kristinas bibliotek i Rom. (Résumé: La bibliothéque de la Reine Christine de Suède à Rome.) In: *Nordisk tidskrift för bok- och biblioteksväsen 1950,* p. 114–124

11. Etudes sur le cadastre médiéval en Suisse romande. In : *Revue d'histoire suisse* 1950, p. 198–243

1951

12. *De stående skatterna på jord 1400–1600.* (Summary: Annual land-taxes in Sweden during the fifteenth and sixteenth centuries.) Lund 1951, 437 p.
13. Folkunganamnet. In: *Arkiv för nordisk filologi* 1951, p. 106–109
14. Les méthodes de l'histoire agraire. In : *Annales. Économies – Sociétés – Civilisations* 1951, p. 340–344
15. "Mark karlgill". Ett bidrag till det fornsvenska bötessystemets historia. In: *Rättshistoriska studier* 1951, p. 184–188
16. Nouvelles recherches sur la bibliothéque de Grotius en Suède et en Italie. In: *Mededelingen de Koninklijke Nederlandse Akademie van wettenschappen, Afd. letterkunde,* N.R. 1951, p. 331–338
17. Review of: Gerhard Hafström, Ledung och marklandsindelning (Uppsala 1949). In: *Rättshistoriska studier* 1951, p. 261–266

1952

18. *Hugo de Groot, Inleidinge tot de Hollandsche rechts-geleerdheid, Met de te Lund teruggevonden verbeteringen, aanvullingen en opmerkingen van den schrijver en met verwijzingen naar zijn andere geschriften uitgegeven en van aantekeningen en bijlagen voorzien,* eds. F. Dovring, H.F.W.D. Fischer en E. M. Meijers. (Preface in Dutch and English.) Leiden 1952 p. xxx+400 (Later editions, see no. 73, 89, 128)
19. Contribution á l'étude de l'organisation des villages Normands au Moyen Age. In: *Annales de Normandie* 1952, p. 99–123, 1 plate
20. Le probléme du manse le Sundgau. Etude d´histoire agraire d'après les documents cadastraux de l'abbaye de Saint-Alban de Bâle. In: *L'Alsace et la Suisse à traverse les siècles* 1952, p. 43–53
21. *Oförrättat uppdrag? Erinringar mot de sakkunnigas utlåtanden om de sökande till den lediga professuren i historia vid Uppsala universitet.* Lund 1952, p. 23

1953

22. *Agrarhistorien. En översikt av dess uppgifter, forskningsmetoder och resultat.* Stockholm 1953, p. 128 (Later edition, no. 75)
23. Hugo Grotius at work on Dutch civil law at Loevestein and in Paris. In: *Atti del terzo congresso di diritto comparator* 2, Rome 1953, p. 305–308
24. Kring jordreformen i Italien. In: *Samtid och framtid* 1953, p. 248–254
25. Ur storsocknarnas historia. In: *Luleå stift i ord och bild.* Stockholm 1953, p. 81–100
26. Vad gör Franco och Salazar åt jordfrågan? In: *Samtid och framtid* 1953, p. 444–450
27. Review of: Jordebok för Värmland 1540, med inledning och kommentarer utgiven av Richard Broberg (Uppsala 1952). In: *Historisk tidskrift* 1953, p. 366–367

28. Review of: Zeitschrift für Agrargeschichte und Agrarsoziologie 1953. In: *Historisk tidskrift* 1953, p. 280–281

1954

29. Agrarhistorisk forskning och svensk medeltidshistoria. In: *Historisk tidskrift* 1953 (issued 1954), p. 384–409
30. La convervazione dell'azienda agraria come unitá giuridica. In: *Rivista di diritto agrario* 1954, p. 224–232
31. Frankrikes bönder vid skiljovägen. In: *Samtid och framtid* 1954, p. 107–114
32. Review of: Björn Helmfrid, Ståthöga norrköpingsgård (Linköping 1952). In: *Historisk tidskrift* 1954, p. 302–303

1955

33. Englands storgods möter nutiden. In: *Samtid och framtid* 1955, p. 100–104
34. Les recensements agricoles français. In: *Bulletin mensuel de statistique, Supplément trimestriel (Etudes statistiques)* 1955, p. 70–74
35. Betriebsgrösse und Arbeitsbesatz. In: *Agrarwirtschaft* 1955, p. 100–105
36. Review of: E. Bertell, Kyrkans och kronans jord på Åland under 1500-talet (Mariehamn 1953). In: *Historisk tidskrift* 1955, p. 80–81
37. Review of: Jörgen Weibull, Tionden i Skåne under senare delen av 1600-talet. In: *Historisk tidskrift* 1955, p. 341–346

1956

38. Attung. In: *Kulturhistoriskt lexikon för nordisk medeltid* 1, col:s 276–278
39. *Land and labor in Europe 1900–1950. A comparative survey of recent agrarian history.* (With a chapter on Land reform as a propaganda theme by Karin Dovring) The Hague 1956, viii+480 p. (New editions, no. 46, 67)
40. Review of: Fridlev Skrubbeltrang, Kyösti Haataja, Pálni Einarsson, Johan Teigland & Stig G:son Berg, Jordpolitiken i Norden (Stockholm 1955). In: *Historisk tidskrift* 1956, p. 68–69
41. Review of: S.D. Zagoroff, Jenö Vegh & Alexander D. Bilimovich, The agricultural economy of the Danubian countries 1935–45 (Stanford, Cal. 1955). In: *Economic History Review* 1956, p. 388–389

1959

42. The share of agriculture in a growing population. In: *Monthly bulletin of agriculture economics and statistics* (FAO) 1959 Vol. 8–9, p. 1–11 (Translations, no. 43, 44, 47, 63, 91, 93, 103, 173)
43. El papel de la agricultura dentro de las poblaciones en crecimiento. In: *Boletin mensual de estadistica y economia agricolas* (FAO) 1959 Vol. 8–9, p. 1–11 (Spanish version of no. 42)
44. La part de l'agriculture dans une population en voie d'accroissement. In: *Bulletin mensuel de statistique et d'économie agricoles* (FAO) 1959 Vol. 8–9, p. 1–11 (French version of no. 42)

1960

45. *History as a social science. An essay on the nature and purpose of historical studies.* The Hague 1960, 97 p. (New edition no. 222, Japanese translation, no. 127)
46. *Land and labor in Europe 1900–1950. A comparative survey of recent agrarian history.* (With a chapter on Land reform as a propaganda theme by Karin Dovring.) The Hague 1960, viii+480 p. (Second edition of no. 39, with some corrections)
47. Pro diminuire rapidamente de la populazione agricola? In: *Mercurio* 1960, p. 23–28 (Digest of no. 44)

1961

48. Review of: H. Cecil Pawson, Cockle Park Farm. An account of the Cockle Park experimental station from 1896 to 1956 (London 1960). In: *Agricultural History* 1961, p. 50

1962

49. European reactions to the Homestead Act. In: *Journal of Economic History* 1962, p. 461–472
50. *Farm size data: Frequency distribution, interpolation, and projection.* Department of Agricultural Economics, University of Illinois 1962 (mimeo), 14 p. (Re-issued 1964)
51. Flexibility and security in agrarian reform programs. In: *Agrarian reform & economic growth in developing countries. Papers from a seminar on research perspectives and problems.* Washington (United States department of Agriculture) 1962, p. 30–40
52. With Hans K. Larsen. The Danish farm system. In: *Illinois Agricultural Economics* July 1962, p. 23–31
53. The opportunity to multiply: Demographic aspects of modern colonialism. In: *Journal of Economic History* 1961, p. 599–612
54. Peasantry, land use and change. Review of Walter Firery, Man, mind and land (Glencoe 1960). In: *Comparative Studies in Society and History* 1962, p. 364–374

1963

55. With William H. Scofield. *Farm real estate sales in Illinois.* Department of Agricultural Economics, University of Illinois 1963, 48 p. (Re-issued 1964)
56. *Labor used for agricultural production. An attempt at a fresh approach to productivity in agriculture.* Department of Agricultural Economics, University of Illinois 1963 (mimeo), 9 p.
57. Zur Bodenreformfrage in unterentwickelten Ländern. In: *Agrarwirtschaft* 1963, p. 221–228

1964

58. Discussion (on Ph. Raup, Public land-use controls in Europe). In: *Land use controls,* Urbana (University of Illinois) 1964, p. 34–35

59. Labor force composition and economic structure of the agricultural sector. In: *Illinois Agricultural Economics* Dec. 1964, p. 29–38
60. Land ownership and tenure reforms. In: *Illinois Agricultural Economics* Dec. 1964, p. 94–102
61. Manpower problems in agriculture. In: *FATIS (Food and Agriculture Technical Information Service) Review* (OECD) 1964:1, p. 5–6 (French translation, no. 62)
62. Problémes de main-d'œuvre agricole. In: *FATIS, Revue International de développment agricole* (OECD) 1964:1, p. 5–6 (French translation of no. 61)
63. The share of agriculture in a growing population. In: *Agriculture in economic development,* eds. Carl Eicher & Lawrence Witt. New York 1964, p. 78–98 (Reprinted with corrections from no. 42, Spanish translation from this version in no. 93)
64. Review of: Johan Hvidtfeldt, Kampen om ophaevelsen af livegenskabet i Slesvig og Holsten, 1795–1805 (Copenhagen 1963). In: *Agricultural History* 1964, 9. 184–185 (Cf. no. 65)

1965

65. Bondage, tenure and progress. Reflections on the economics of forced labor: Reviews of: I.M. Confino, Domaines et seigneurs en Russie vers la fin du 18 siècle (Paris 1963); J. Hvidtfeldt, Kampen om ophaevelsen af livegenskabet i Slesvig og Holsten, 1795–1805 (Copenhagen 1963); C. Poni, Gli aratri e l´economia agrarian nel Bolognese dal 17 al 19 secolo (Bologna 1963). In: *Comparative Studies in Society and History* 1965, p. 309–323
66. *Estimating cross tables. A study on frequency distributions of economic data classified by size.* Department of Agricultural Economics, University of Illinois 1965 (mimeo), 20 p.
67. *Land and labor in Europe in the Twentieth Century. A comparative survey of recent agrarian history.* (With a chapter on Land Reform as a propaganda theme by Karin Dovring). The Hague 1965, xi+511 p. (Third revised ed. of no. 39, cf. no. 46)
68. *Problems of manpower in agriculture,* Paris (OECD)1965, 158 p. (French version, no. 69)
69. *Problémes de la main-*d'œuvre agricole, Paris (OECD) 1965, 170 p. (French translation of no. 68)
70. The transformation of European agriculture. In: *The Cambridge economic history of Europe* 6, Cambridge 1965, p. 603–672, 1007–1011 (Italian translation, no. 144)
71. *Condemnation, Land Value, and Severance Damage on Farmland.* University of Illinois, Agricultural Experiment Station Bulletin 707 1965, 93 p. (Dovring main author, see p. 3)
72. When farmers become fewer. In: *Illinois Agricultural Economics* July 1965, p. 1–9 (Cf. no. 80, 86)

73. *Hugo de Groot, Inleidinge tot de Hollandsche rechts-geleerdheid,* eds. F. Dovring, H.F.W.D. Fischer en E.M. Meijers. (Preface in Dutch and English.) Leiden 1965, p. xxx+400 (Second edition of no. 18, with corrections)
74. Review of: Slicher van Bath, The agrarian history of Western Europe A.D. 500–1850 (New York 1963). In: *Journal of Economic History* 1965, p. 295–297

1966

75. *Agrarhistorien.* Stockholm 1966, 128 p. (Reprint of no. 22)
76. *Farmland prices increases in Illinois 1954–1964. Differences by value level.* Department of Agricultural Economics, University of Illinois 1966 (mimeo), 4 p.
77. With William Gossling. Labor productivity measurement. The use of subsystems in the interindustry approach, and some approximating alternatives. In: *Journal of Farm Economics* 1966, p. 369–377
78. *Land reform and productivity. The Mexican case, a preliminary analysis.* Department of Agricultural Economics, University of Illinois 1966 (mimeo), 17 p. (Cf. no. 101, 111, 113, 117, 122)
79. Soviet farm mechanization in perspective. In: *Slavic Review* 1966, p. 287–302 (Cf. no. 83)
80. When farmers become fewer. In: *Agricultural Policy Review* 1966, p. 8–10 (Digest of no. 72)
81. Review of: Ester Boserup, The conditions of agricultural growth. The economics of agrarian change under population pressure (Chicago 1965). In: *Journal of Economic History* 1966, p. 380–381
82. Review of: Soviet agriculture. The permanent crisis, ed. Roy D. Laird (New York & London 1965). In: *Slavic Review* 1966, p. 551

1967

83. Meccanizzazione e colettivismo nell´agricoltura sovietica. In: *Mercurio* 1967, p. 3–7 (Digest of no. 79)
84. *Productivity of labor in agricultural production.* University of Illinois, Agricultural Experiment Station Bulletin 726 1967, 73 p.
85. Underemployment in traditional agriculture. In: *Economic Development and Cultural Change* 1967, p. 163–173 (The title erroneously "Unemployment", Errata in next issue. Sequel, no. 105)
86. Wenn es weniger Bauern werden. In: *Agrarwirtschaft* 1976, p. 87–93 (German translation of no. 72)
87. Review of: William C. Thiesenhusen, Chile's experiments in agrarian reform (Madison, Wisconsin 1966). In: *Journal of Farm Economics* 1967, p. 538–540
88. Review of: Elias H. Tuma, Twenty-six centuries of agrarian reform. A comparative analysis (Berkeley & Los Angeles 1965). In: *Political Science Quarterly* 1967, p. 280–281

1968

89. *Hugo de Groot, Inleidinge tot de Hollandsche rechts-geleerdheid,* eds. F. Dovring, H.F.W.D. Fischer en E.M. Meijers. (Preface in Dutch and English.) Leiden 1965, p. xxx+400 (Reprint of no.73, see no. 18, with new corrections)
90. A new dimension to "surplus" disposal. Aid by dumping? In: *International Development Review* Sept. 1968, p. 2–5 (Cf. correction of bibliography in *International Development Review* 1968:1, Italian digest in no. 100)
91. El papel de la agricultura dentro de las poblaciones en crecimiento. Mexico, un caso de desarrollo económico reciente. In: *El Trimeste Economico* 1968, p. 25–50 (Reprinted from no. 43, p. 47–50 added for this version)
92. *Income growth rate and sector proportions. The share of agriculture at successive levels of income.* Department of Agricultural Economics, University of Illinois 1968 (mimeo), 10 p.
93. La parte decidada a la agricultura de un población creciente. In: *La agricultura en el desarollo economico*, ed. Carl K. Eicher & Lawrence W. Witt. Mexico 1968, p. 93–113 (Spanish translation of no. 63, cf. no. 42)
94. Review of: Agriculture and economic growth in England 1650–1815, ed. E.C. Jones (London 1967). In: *The Social Studies* 1968, p. 294
95. Reviews of: Gilbert Etienne, Studies in Indian agriculture. The art of the possible (Berkeley & Los Angeles 1968); Tadashi Fukutake, Asian rural society. China, India, Japan (Seattle & London 1967). In: *Journal of Economic History* 1968, p. 467–469
96. Review of: Barrington Moore, Jr., Social origins of dictatorship and democracy. Lord and peasant in the making of the modern world (Boston 1966). In: *Labor History* 1968, p. 141–144

1969

97. The principle of acceleration. A non-dialectical theory of progress. In: *Comparative Studies in Society and History* 1969, p. 413–425 (Cf. no. 98)
98. Counter-comment (reply to Th. S. Kuhn on The principle of acceleration no. 97). In: *Comparative Studies in Society and History* 1969, p. 431–432
99. Review of: R.A. Rappaport, The case for a static society. Pigs for ancestors. Ritual in the ecology of a New Guinea people (New Haven 1967). In: *Comparative Studies in Society and History* 1969, p. 227–233
100. "Dumping" e aiuti all'estero. In. *Mercurio* 1969, p. 57–62 (Digest of no. 90)
101. Eighteenth-century changes in European agriculture. A comment (On Slicher van Bath). In: *Agricultural History* 1969, p. 181–186
102. *Land reform and productivity. The Mexican case. Analysis of census data.* Madison, Wisconsin (University of Wisconsin, Land Tenure Centre Paper 63) 1969 (mimeo), 22 p. (Revised version of no. 78, Spanish translation no. 114)
103. Der landwirtschaftilche Anteil an einer wachsenden Bevölkerung. In: *Agrarwirtschaft und Agrarpolitik.* Hrsg. Eberhard Gerhardt & Paul Kuhlman. Köln & Berlin 1969, p. 346–366 (Translation of no. 42, diagrams updated)

104. *Income and wealth distributions. The exponential functions.* Department of Agricultural Economics, University of Illinois 1969 (mimeo), 21 p.
105. Rejoinder (to A.A. Pepelasis, on underemployment in traditional agriculture). In: *Economic Development and Cultural Change* 1969, p. 273–276 (Cf. no. 85)
106. Review of: Robert Conquest. Agricultural workers in the U.S.S.R. (New York & Washington 1968). In: *Journal of Economic History* 1969, p. 779–780
107. Review of: Rainer Schickele, Agrarian revolution and economic progress. A primer for development (New York 1968). In: *Land Economics* 1969, p. 125–128

1970

108. *Economic results of land reforms*, Washington D.C. (Agency for International Development, A.I.D) 1970 (mimeo), 28 p. (Cf. 109–113)
109. Economic results of land reform. In: *War on hunger*, Washington D.C (A.I.D.) 1970, p. 12–15
110. *Land reform in Hungary*. Washington D.C. (A.I.D.) 1970 (mimeo), 59 p.
111. *Land reform in Mexico*. Washington D.C. (A.I.D.) 1970 (mimeo), 66 p.
112. *Land reform in Yugoslavia*. Washington D.C. (A.I.D.) 1970 (mimeo), 61 p.
113. Summary paper on Europe. In: *Land reform regional surveys*. Washington D.C. (A.I.D.) 1970 (mimeo), p. 1–5
114. *Land Economics. A syllabus for classroom treatment.* Department of Agricultural Economics, University of Illinois 1970 (mimeo), 108 p. (Revised edition no. 176)
115. Land reform and productivity in Mexico. In: *Land Economics* 1970, p. 264–274 (Italian digest in no. 122)
116. Letters to the editor. Lenin and Elisabeth K. In: *Slavic Review* 1970, p. 570–573
117. Reforma agraria y productividad. El caso Mexicano. Análisis de datos censales. In : *Investigación económica* 1968 (issued 1970), p. 167–168 (Translation of no. 102, reprinted in 140)
118. Variants and invariants in comparative agricultural systems. In: *American Journal of Agricultural Economics* 1969 (issued 1970), p. 1263–1273
119. Review of: S.H. Franklin, The European peasantry, the final phase (London 1969). In: *Journal of Economic History* 1970, p. 457–459

1971

120. *Farmland prices increases in Illinois 1954–1969. Differences by price level. Department* of Agricultural economics, University of Illinois 1971 (mimeo), 3 p. (Revision of no. 76)
121. Progress of mechanization in Soviet agriculture. In: *The Soviet Rural Community*, ed. James R. Millar. Urbana, Illinois, 1970, p. 259–275
122. Riforma agraria e produttivitá nel Messico. In: *Mercurio* 1971, p. 88–91 (Digest of no. 111)

123. Review of: Agricultural policy in an affluent society, eds. Vernon W. Ruttan, Arley D. Waldo, James P. Houck (New York 1969). In: *The Social Studies* 1971, p. 28

124. Review of: Agrarian change and economic development. The historical problems, eds. E.L. Jones & S.J. Woolf (New York & London 1969). In: *American Journal of Agricultural Economics* 1971, p. 145–146

125. Review of: Henri Mendras, The vanishing peasant. Innovation and change in French agriculture (Cambridge, Mass. & London 1970). In: *Journal of Economic History* 1971, p. 979–981

1972

126. With Karin Dovring. *The optional society. An essay on economic choice and bargains of communication in an affluent world.* The Hague 1971 (issued 1972), x+107 p.

127. *Rekishi to wa nani ka? Shakai kagaku toshite no rekishigaku.* Kyoto 1972, 150 p. (Japanese translation of no. 45, *History as a social science*)

128. *Hugo de Groot, Inleidinge tot de Hollandsche rechts-geleerdheid,* eds. F. Dovring, H.F.W.D. Fischer en E.M. Meijers. (Preface in Dutch and English.) Leiden 1972 (Paperback edition of no. 73, 89)

129. Macro constraints on agricultural development in India. In: *Indian Journal of Agricultural Economics* 1972, p. 46–66

130. With J.R. Jindia & R.S. Misra. *Economic production possibilities for soybeans in Northern India. A preliminary study.* Urbana-Champaign, Illinois (University of Illinois Office of International Programs and Studies) 1972, 62 p.

131. Review of Colin Clark, Starvation or plenty? (New York 1970). In: *The Social Studies* 1972, p. 45–46

132. Review of: Edith Raskin, World food (New York 1971). In: *The Social Studies* 1972, p. 342–343

1973

133. Iskusstvennoe miâsa iz soi. In: *Voprosy Ékonomiki* 1973, p. 92–94 (Cf. no. 136)

134. Distribution of farm size and income. Analysis by exponential functions. In: *Land Economics* 1973, p. 133–147

135. With H.G. Halcrow and others. *School tax options affecting Illinois agriculture,* University of Illinois, Agricultural Experiment Station Bulletin 744, 1973, 35 p.

136. Artificial meat from soybeans. In: *OECD Agricultural Review* 1973, p. 122–123 (Same as no. 133, OECD also issued the text in French, German, Greek, Spanish and Turkish. French text reprinted in: *Agriculture. Revue mensuelle,* (Paris), May 1974, p. 128–129)

137. Review of: Agrarian policies and problems in Communist and non-communist countries, ed. W.A. Douglas Jackson (Seattle & London 1971). In: *Political Science Quarterly* 88:1, 1973, p. 117–119

138. Review of: Barbara Lewis Solow, The land question and the Irish economy 1870–1903 (Cambridge, Mass. 1971). In: *Political Science Quarterly* 1973, p. 518–519

1974

139. Soybeans. In: *Scientific American*, Feb 1974, p. Cover + 14–21 (Reprinted in *Congressional Record* March 11 1974, p. 3295–3 297)

140. Reforma agraria y productividad. El caso Mexicano. In: *La Economia Mexicana 1. Análisis por sectores de distribución*. Selección de Leopoldo Solls. México 1973 (issued 1974), p. 35–45 (Reprint of no. 117)

141. With R.M. Leuthold & G.L. Karr. Distributive equity of incomes in rural Illinois. In: *Rural community and regional development. Perspectives and prospects*, ed. John T. Scott. Department of Agricultural Economics, University of Illinois 1974, p. 41–52

142. *The world food crisis and the challenge to agriculture.* Chapel Hill, North Carolina (Institute of Nutrition, The university of North Carolina), 1974, 14 p. (The inaugural H. Brooks James Memorial Lecture March 26 1974)

143. Land reform. A key to change in agriculture. In: *Agricultural policy in developing countries*, ed. Nurul Islam. London 1974, p. 509–521

144. Le trasformazione dell'agricoltura europea. In: *Storia economica Cambridge 6. La rivoluzione industriale a i suoi sviluppi.* Torino 1974, p. 653–730 (Italian translation of no. 70)

145. Inflation feeds on gasoline. In: *Illinois Business Review*, Oct 1974, p. 6–8 (Reprinted in *Congressional Record* March 4 1974, p. E 878–879)

146. Review of: Antony C. Sutton, Western technology and Soviet economic development 1945 to 1965 (Stanford. Cal. 1973). In: *Soviet Union – Union Soviétique* 1974, p. 201–202

147. Review of: Economic models and quantitative methods for decision and planning in agriculture, ed. E.O. Heady (Ames, Iowa 1971). In: *Canadian-American Slavic Studies* 1972, p. 613–614

1975

148. The conditions for unlimited economic growth. In: *Planning and Public Policy* 1:1 (Bureau of Urban and Regional Planning, University of Illinois) 1974–975, p. 1–4

149. Population and food. The development problem. In: *The Great Lakes Assembly on Population and Hunger*. East Lansing, Michigan (Kellogg Center, Michigan State University) 1975, p. 13–19

150. *A national economic policy for the United States.* Urbana, Illinois (published by the author) 1975 (mimeo), 16 p.

151. Review of: Environmental deterioration in the Soviet Union and Eastern Europe, ed. Ivan Volgyes (New York 1974). In: *Russian Review* 1975, p. 219–220

152. Review of: Assar Lindbeck, Swedish economic policy (Berkeley & Los Angeles 1974). In: *Journal of Economic History* 1975, p. 884–885

1976

153. The State of Illinois and international programs of the College of Agriculture. In: *Current Affairs of the College of Agriculture*, Urbana, Illinois (University of Illinois) Jan.–Feb. 1976, p. 3–4

154. Economic impact of foreign investment in real estate. In: *Foreign Investment in U.S. Real Estate*. Washington D. C. (United States Department of Agriculture, Economic Research Service) 1976, p. 132–146

155. Economic impact of foreign investment in real estate. In: *Foreign direct investment in the United States. Report of the Secretary of Commerce to the Congress in compliance with the Foreign Investment Study Act of 1974*. Washington D.C. (United States Department of Commerce) 1976, p. L–75 (Summary of no. 154, also in Gene Wunderlich, Summary of the report: Foreign investment in the U.S. real estate, *USDA Agricultural Information Bulletin* no. 400 1976, p. 21)

156. Energy. A clear and present danger. In: *The Washington Post* Oct. 30 1976

157. Party platforms blueprints for disaster. In: *Illinois Business Review* Oct. 1976, p. 2, 12

158. Review of: Lewis. A. Fisher & Philip E. Uren, The new Hungarian agriculture (Montreal & London 1973). In: *Canadian-American Slavic Studies* 1976, p. 156

159. Review of: Karl-Eugen Wädekin, Sozialistische Agrarpolitik in Osteuropa 1 (Berlin 1974). In: *Slavic Review* 1976, p. 358–369

1977

160. The farmland boom in Illinois. In: *Illinois Agricultural Economics* July 1977, p. 34–38

161. *Gross and net productivity. A problem of aggregation level.* Department of Agricultural Economics, University of Illinois (mimeo), 1977, 8 p.

162. *Transnational business, its promise and problems.* Champaign, Illinois (University YMCA) (mimeo), 1977, 9 p.

163. Energy doubling by year 2000? Critique of a growth perspective. In: *Congressional record* Dec. 6 1977, p. S 19354–S 19356

164. Review of: Mechanisation and employment in agriculture. Case studies from four continents (Geneva 1973). In: *Economic Development and Cultural Change* 1977, p. 764–769

165. Review of: Heinrich Schweizer, Sozialistische Agrartheorie und Landwirtschaftspolitik in China und der Sowjetunion. Ein Modell für Entwicklungsländer? (Bern & Frankfurt/M 1972). In: *Canadian-American Slavic Studies* 1976 (issued 1977), p. 621

166. Review of: Maria de Jong Ellis, Agriculture and the State in ancient Mesopotamia. An introduction to problems of land tenure (Philadelphia 1976). In: *Journal of Economic History* 1977, p. 1054–1055

167. Review of: Jack Goody, Production and reproduction. A comparative study of the domestic domain (Cambridge 1976). In: *Journal of Economic History* 1977, p. 1064–1065

1978

168. Drive to oblivion. Gas and the wasting cities. In: *The Morning Courier* (Champaign-Urbana) April 19 1978

169. Declare war on energy problem. In: *The Mourning Courier* (Champaign-Urbana) May 28 1978

170. The farm land boom in perspective. Recent history and economic relationships of farmland values. In: *Effects of high farm land prices.* Department of Agricultural Economics, University of Illinois July 1978, p. 1–11

171. Macro-economics of farm mechanization. In: *Agricultural technology for developing nations. Farm mechanization alternatives for 1–10 hectare farms.* Urbana, Illinois (University of Illinois) 1978, p. 16–19 (Reprint see no. 199)

172. Scandinavia: Denmark, Finland, Iceland, Norway, Sweden. In: *Crisis of political development in Europe and the United States*, ed. R. Grew. Princeton 1978, p. 139–162

173. The share of agriculture in a growing population. In: *FAO Studies in Agricultural Economics and Statistics 1952–1977.* Rome 1978, p. 186–196 (Reprint of no. 42)

1979

174. Cropland reserve for fuel production. *A report for the alcohol from biomass task study.* Department of Agricultural Economics, University of Illinois 1979 (mimeo), 21 p. (Also issued in: *Raw material availability report*, Sept. 1979, National Technical Information Service)

175. Underemployment, slow motion and X-efficiency. In: *Economic Development and Cultural Change* 1979, p. 485–490

176. *Land economics. A syllabus for classroom treatment.* Department of Agricultural Economics, University of Illinois 1979 (mimeo), 130 p. (Revised edition of no. 114, printed version no. 233)

177. With John F. Yanagida. *Monoculture and productivity. A study of private profit and social product on grain farms and livestock farms in Illinois.* Department of Agricultural Economics, University of Illinois 1979, 24 p.

178. *The two faces of economic distribution. Population and distributed good.* Department of Agricultural Economics, University of Illinois 1979 (mimeo), 14 p.

179. Gross and net productivity. A problem of aggregation. In: *American Journal of Agricultural Economics* 1979, p. 694–696

180. Development as history. Review of: Jerome Blum, The end of the old order in rural Europe (Princeton 1978); Shigemochi Hirashima, The structure of disparity in developing agriculture. A case study of the Pakistan Punjab (Tokyo 1978). In: *Comparative Studies in Society and History* 1979, p. 567–571

181. Review of: Irving B. Kravis, Alan Heston & Robert Summers, International comparisons of real product and purchasing power (Baltimore 1978). In: *American Journal of Agricultural Economics* 1979, p. 385–387

1980

182. With Donald R. McDowell. *Energy used for fertilizers*. Department of Agricultural Economics, University of Illinois 1980 (mimeo), 20 p.
183. *Export or burn? American grain and the energy equations*. Department of Agricultural Economics, University of Illinois 1980 (mimeo), 23 p.
184. Export or burn? American grain and the energy equations. In: *Illinois Business Review* May 1980, p. 9–12 (Summary of no. 183)
185. Capital intensity in Soviet agriculture. In: *Agricultural Policies in the USSR and Eastern Europe*, eds. Ronald A. Francisco, Betty A. Laird & Roy D. Laird. Boulder. Colorado 1980, p. 5–26
186. Effects of fermentation alcohol on the agricultural economy. In: *The First Inter American Conference on Renewable Sources of Energy*. New Orleans, Louisiana 1980, p. 245–247
187. With H.G. Halcrow and others. *Illinois farm property taxes*. Department of Agricultural Economics, University of Illinois 1979 (issued 1980), 28 p.
188. Soviet agriculture in 1980. In: *Current History* 1980, p. 88–91, 105–106
189. With John B. Braden. *Do we need to be concerned about farmland?* Department of Agricultural Economics, University of Illinois 1980 (mimeo), 6 p.
190. With Donald R. McDowell. *Energy used for pesticides*. Department of Agricultural Economics, University of Illinois 1980 (mimeo), 11 p.
191. With Robert A. Herendeen, Randall Plant & Mary Anne Ross. *Fuel alcohol from grain. Energy and dollar balances of small ethanol distillers and their economies of size and scale*. Department of Agricultural Economics, University of Illinois 1980 (mimeo), 60 p.
192. Review of: Russel King, Land reform. A world survey (Boulder, Colorado 1977). In: *Economic Development and Cultural Change* 1980, p. 878–882
193. Review of: The political economy of collectivized agriculture. Comparative study of communist and non-communist systems, ed. Ronald A. Francisco, Betty A. Laird & Roy D. Laird (New York & Oxford 1979). In: *Slavic Review* 1980, p. 513–514

1981

194. Transportation fuels inflation. In: *Illinois Business Review* February 1981, p. 8–10, 12
195. The oil crisis. A case of American underdevelopment. In: *Illinois Business Review* April 1981, p. 1–3, 7, 11
196. Oil crisis and wishful thinking. In: *Illinois Business Review* Oct. 1981, p. 9–10
197. *Land use in Champaign County, Illinois*. Department of Agricultural Economics, University of Illinois 1981 (mimeo), 10 p.

198. Energy production and natural resource trade-offs. In: *Economics, Ethics, Ecology. Roots of productive conservation*, ed. Walter E. Jeska. Ankeny, Iowa 1981, p. 353–363
199. Macro-economics of farm mechanization. In: *Culture and Agriculture, Bulletin of the Anthropological study group on agrarian systems* (Urbana, Illinois) 1981, p. 5–9 (Reprinted from no. 171)
200. *Energy used for processed livestock feed.* Department of Agricultural Economics, University of Illinois 1981 (mimeo), 15 p.
201. Review of: Wilhelm Abel, Agricultural fluctuations in Europe from the thirteenth to the twentieth centuries (London & New York 1980). In: *Land economics* 1981, p. 296–300

1982

202. *Energy analysis and economic forecasting.* Department of Agricultural Economics, University of Illinois 1982 (mimeo), 22 p.
203. *Area measure in Illinois.* Department of Agricultural Economics, University of Illinois 1982 (mimeo), 9 p.
204. With John B. Braden & David Chicoine. Farmland conversion and preservation in Illinois. In: *Illinois CRD report* (University of Illinois) March 1982, p. 1–5
205. Oil crisis and vested interest. In: *Illinois Business Review* April 1982, p. 10–11
206. Effects of an alcohol fuels program on farm economics. In: *Agriculture as a producer and consumer of energy*, ed. William Locker, Boulder, Colorado, 1982, p. 41–60
207. *Land potential for biomass production in Illinois.* Department of Agricultural Economics, University of Illinois 1982 (mimeo), 25 p.
208. New plans and old results for Soviet agriculture. In: *Current History* 1982, p. 323–326, 341–342
209. *Concepts of land, value and wealth.* Department of Agricultural Economics, University of Illinois 1982 (mimeo), 19 p.
210. With David Chicoine & John Braden. Evaluating agricultural land use changes in Illinois. In: *Journal of Soil and Water Conservation* 1982, p. 350–361

1983

211. Area measurements in Illinois. In: *Transactions of Illinois Academy of Sciences* 1982 (issued 1983), p. 239–246
212. Concepts of land, value and wealth. In: *Land – Something of value*, ed. Gene Wunderlich. Cambridge, Mass. 1982 (issued 1983), p. 11–19
213. Review of: The modernization of agriculture. Rural transformation in Hungary, 1848-1975, ed. Joseph Held (Boulder, Colorado 1980). In: *Canadian-American Slavic Studies* 1982 (issued 1983), p. 297–298
214. Review of: Karl-Eugen Wädekin, Agrarian policies in communist Europe. A critical introduction (Totowa & The Hague 1982). In: *Slavic Review* 1983, p. 499–500

1984

215. *Energy use in United States agriculture. A critique of recent research.* Department of Agricultural Economics, University of Illinois 1984 (mimeo), 13 p.
216. *Riches to rags. The political economy of social waste.* Cambridge, Mass. 1984, ix+171 p.
217. Interest rates, a form of inflation? In: *Illinois Business Review* April 1984, p. 4–5, 12
218. *The cost of agricultural growth and development.* Department of Agricultural Economics, University of Illinois 1984 (mimeo), 16 p.
219. Soviet agriculture – a state secret. In: *Current History* 1984, p. 323–326, 338–339
220. *Energy use for Midwest agriculture.* Department of Agricultural Economics, University of Illinois 1984, 56 p.
221. With Robert A. Herendeen. *Liquid fuel from Illinois sources.* Department of Agricultural Economics, University of Illinois 1984, 61 p.

1985

222. *History as a social science.* New York & London 1984 (issued 1985), 97 p. (Reprint of no. 45)
223. Energy use in the United States agriculture. A critique of recent research. In: *Energy in Agriculture* 1985, p. 79–86
224. Greyhounds racing proponents' arguments on benefits fallacious. In: *The Champaign-Urbana News-Gazette* May 17 1985, p. 4–5
225. Economic arguments against dog racing. In: *Illinois Standarbred and Sulky News* June 1985, p. 52–54 (Reprint of 224)
226. *Gross sales as a measure of farm size.* Department of Agricultural Economics, University of Illinois 1985 (mimeo), 17 p.
227. *Marginal cost of exported crops. The case of corn.* Department of Agricultural Economics, University of Illinois 1985 (mimeo), 19 p.
228. Review of: Land reform, American style, eds. Charles Geisler & Frank J. Popper (Totowa 1984). In: *Rural Development Perspectives* (USDA) Oct. 1985, p. 44–45

1986

229. With J. J. Jokela. Shelterbelts in the Prairie Province of Illinois. *International Symposium on Windbreak Technology. Proceedings.* Lincoln, Nebraska (Great Plains Agricultural Council) 1986, p. 127–129
230. Review of: Pekka Sutela, Socialism, planning and optimality. A study in Soviet economic thought (Helsinki 1985). In: *Slavic Review* 1985 (issued 1986), p. 736–737

1987

231. *Marginal cost of export crops. Wheat, sorghum and rice.* Department of Agricultural Economics, University of Illinois 1987 (mimeo), 23 p.

232. *Marginal cost of export crops. Soybeans, cotton and tobacco.* Department of Agricultural Economics, University of Illinois 1987 (mimeo), 10 p.
233. *Land economics.* Boston 1987, 532 p. (Printed revised version of no. 176)
234. *Productivity and value. The political economy of measuring progress.* New York 1987, vii+194 p.
235. By faith and credit. Economic growth in the 1980s. In: *Illinois Business Review* Oct. 1987, p. 10–13
236. New directions in Soviet agriculture. In: *Current History* 1987, p. 329–332, 340–341

1988

237. *Progress for food or food for progress? The political economy of agricultural growth and development.* New York 1988, vi+324 p.
238. *Farming for fuel. The political economy of energy resources in the United States.* New York 1988, vi+152 p.
239. Review of: Sustainability issues in agricultural development, ed. Ted J. Davis & Isabelle A. Schirmer (Washington D.C. 1987). In: *The Journal of Agricultural Economics Research* 1988, p. 29–30

1990

240. Cost of agricultural growth and development. A cross-national analysis focusing on the USSR. In: *Soviet agriculture, Comparative perspectives*, ed. Kenneth R. Gray. Ames, Iowa 1986, p. 23–47
241. With Robert A. Herendeen. Biomass. Sustainable energy crop of the future. In: *Agro-Ecology* (University of Illinois) March–April 1990, p. 2–3
242. Alternative to oil can be farmed. In: *S:t Louis Post-Dispatch* Aug. 19

1991

243. *Inequality. The political economy of income distribution.* New York 1991, viii+183 p.

1994

244. Energy farming. A prospect for sustainable agriculture. In: *Sustainable agriculture in the American Midwest. Lessons from the past, prospects for the future*, ed. Gregory McIsaac & William R. Edwards. Urbana & Chicago 1994, p. 231–253

1996

245. *Leninism. Political economy as pseudoscience.* London 1996, xi+155 p.

1998

246. *Knowledge and ignorance. Essays on light and shadows.* London 1998, 176 p.

List of Ph.D. thesis supervised as main advisor

(Year, name of student [first name, second name and family name], title of thesis)

1963, Anthony Michael Grano, Regional factors affecting farmland values in Illinois
1963, Herbert Horst Stoevener, Determination of severe damages resulting from partial taking of farm land
1965, Roger Wallace Strohbehn, Income distribution on selected types of Illinois farms and implications for tenure adjustment
1967, Gottfried Ablasser, A flow-of-funds analysis of the United States farm business sector
1967, Satish Chandra Jha, Indian land reform research literature
1967, Badie Jamil Al-Kaddo, International wheat trade 1950–65 with particular reference to government policies
1967, Donald Alexander Muir, Some aspects of international commodity agreements and the international wheat trade
1970, Ramon Jansuy Cruz, Measuring disguised unemployment by means of labor norms
1970, Jaswant Rai Jindia, Estimation of costs and returns on Indian farms
1972, Matthias Kondrad von Oppen, Optimal size and location of marketing facilities for soybeans in India
1972, Elizabeth Banks Erickson, A linear programming model for analysis of irrigation agriculture in Northwest Mexico
1973, Roop Narayan Pandey, Livestock and crop production patterns in Uttar Pradesh, India
1974, Muhammad Arshad Chawdry, Effect of land tenure on resource use and productivity in agriculture. A case study of Punjab, Pakistan
1977, Richard Haines Bernsten, Constraints to higher rice yields in the Philippines
1978, John Fumio Yanagida, Productivity analysis on a sample of Illinois cash grain farms 1972–1976
1979, David Lyle Chicoine, Farmland values in an urban fringe. An analysis of market data from Will County, Illinois
1980, Mohamoud Awaleh Gulaid, Economic tradeoffs between foodgrain and banana production in Somalia
1981, Fernando Zalacain, A regional and interregional input-output analysis of the Puerto Rican economy
1983, Abbas Ali Kazemi, Economic analysis of modifying cropping patterns in selected regions of Iran
1984, Thomas, Gillard-Byers, The allocation of resources in the non-mechanized small farm households of Dumila, M'Kudzi and Magole villages in the Kiloza district of Tanzania
1985, Donald Ray McDowell, Conventional production functions versus energy-term production functions. Energy used for crop production in Midwest agriculture
1986, Marcia Noreen White, Limited resource countries and agricultural development. A methodology used for the Caribbean

1986, Sadok Driss, Evaluation research and agricultural development. A Tunisian case study

1986, Anthony Boris Soskin, Non-traditional agriculture and economic development. The Brazilian soybean expansion

Dovring also supervised 37 Master's theses, presented 1960–1987

APPENDIX 2

Ranking of universities

Comparing different nations

In the main text I claim that universities in the U.S.A. dominate the university systems of the world. That quite remarkable fact needs to be confirmed with some basic data. Henry Rosovsky assumed in 1990 that the U.S. had two thirds to the three quarters of the best,[392] which is astonishing if correct.

Ranking of universities in the U.S. has been going on for a long time, but until recently has not been of importance in other countries.[393] From the most well known rating in the U.S. today I turn to an international ranking that positions U.S. universities in a wider context. Then I go back to older ratings in the U.S. and especially discuss the persistence of positions.

Most often used today is the rating by *U.S. News and World Report*, which every year since 1983 has produced a national ranking.[394] Its central importance has been the topic of much debate. David Kirp has remarked that "an otherwise little-noticed news magazine" produces a dubious ranking system "that has become higher education's *Michelin Guide*."[395] This evaluation was originally intended for parents and students who were in the process of choosing a school. Besides gathering data on academic excellence, the students' performance and estimation of their school is measured, as are the facilities for students and factors such as class size. Not only the sciences but also the humanities are included in these estimations. The top group tends to be the same: Harvard, Princeton, Yale, MIT, and California Institute of Technology, though the order varies slightly from year to year.[396]

A worldwide survey made by Shanghai Jiao Tong University, China in 2003

392 Rosovsky (1990). He state this fact already in the title of one of his chapters, p. 29–36.

393 Bok (1986), p. 14–15. I will not here go further into research and scientific discussion about institutional rankings, cf. literature in Clark (1987), p. 325. Rating started in the U.S: already in the 1920s and 1930 and a became of importance again from the late 1950s, see Berelson (1960), p. 29, Geiger (1993), p. 211–212.

394 http://www.usnews.com.

395 Kirp (2003), p. 12, cf. p. 268.

396 The top universities in 2000 were, in order: Cal. Tech.; Harvard, MIT, Princeton, and Yale. The highest ranked in 2003 were Princeton; Harvard, Yale, Cal. Tech., Duke, MIT, Stanford, and U. of Pennsylvania; in 2004, Harvard, Princeton, Yale, MIT, Cal. Tech., Duke, Stanford, and Pennsylvania; in 2009, Harvard, Princeton, Yale, MIT, Stanford, and Cal. Tech.

drew immediate attention, and the importance of this rating has tended to grow.[397] Here nothing except research is measured and the rating heavily favors the natural sciences.[398] The weights are: Nobel laureates for faculty and alumni, with less heavy weights for older prizes (economics is included but not literature); highly cited researchers; articles in *Nature* and *Science*; articles included in the database ISI (*Science Citation Index* and *Social Science Citation Index*, but not *Arts and Humanities Citation Index*); these indicators are divided by the full-time staff equivalent. Another list is produced by *Times Higher Education Supplement*,[399] but the Shanghai list dominates the worldwide discussion.

If we look only at American universities an interesting similarity can be shown between the Shanghai list and the one from *U.S. News and World Report*, which uses a partly different set of data and has another goal. Of the top ten in the U.S. on the Shanghai list in 2003, six are among *U.S. News'* top ten, and the rest have numbers 11, 13, 14 and 21 on the *U.S. News* list.[400] In the *Times* list from 2004 we again find the same universities atthe top.[401] The Shanghai list may thus, despite the uncertainties always connected with rankings, be used to estimate the world distribution of top universities.[402]

A source critical aspect is the dominance of the sciences. Universities specializing in the natural sciences (such as technical institutes) often get a higher score than they would have received if a more social science and humanities biased method had been used. But I assume that this does not affect the results on a national level. The United States' dominance in the Shanghai list is not an effect of a general bias in the research toward natural sciences. Of students engaged in tertiary level of education in the U.S. in the 1990s only 14 % were enrolled in natural and applied science curricula. Finland was near the top with 52 %, Sweden had 43 % and Germany 42 %.[403]

---

397 Established by searches on Google: "ranking"+ "universities" from 2003 and until today. See also power point presentations from The 3rd International Symposium on University Rankings Leiden, 6–7 February 2009, see: http://www.events.leiden.edu/ranking2009/presentations.jsp. Here the Shanghai list is the benchmark in many presentations.

398 http://ed.sjtu.edu.cn/ranking.htm.

399 http://www.topuniversities.com/worlduniversityrankings/. Here the Nobel prize is not an indicator; instead peer reviews from a number of scholars are an important factor.

400 The six on both lists are, in the Shanghai order: Harvard, Stanford, Cal. Tech., MIT, Princeton, and Yale. The others on the Shanghai list are in the ranking order on the U.S. News list: Columbia, Chicago, Cornell, and UC/Berkeley. The only real difference is Berkeley being ranked 4 on the Shanghai-list and 21 on the U.S. News list.

401 Of the top ten in the *Times* list in 2004 eight universities are from the U.S., in order: Harvard, Berkeley, MIT, Cal. Tech. – then the *Times* lists Oxford and Cambridge – followed by Stanford, Yale, and Princeton.

402 The top on the Shanghai list has been nearly constant 2003–2008, with Harvard ranked first, Stanford second, and so on.

403 Kurian (1997), p. 304–305. Other countries are, for instance, the United Kingdom 39 %, Australia 33 %, France 31 %, Japan 26 % or Canada 14 %.

Another bias derives from the different university systems. In France much of the best research is done in specific research institutes; many publications are written in French and thus fall outside the evaluation. French university administrators got a chock by the publication of the first Shanghai list, and since then the organization and administration of French research has been changed. Many of the research institutes have now been incorporated into universities.

In table 4 I have used the ranking from the Shanghai list in 2003, including every country that has a university among the top 100, and counted the 300 best universities in the world. The order is arranged so that the countries' position among the first 50 is highlighted. These are the "gold medals" that determine the order. Then I include the position among the first 100, "gold and silver medals," and so on. The reason for organizing the table thus is to emphasize the top positions.

An important result is that United States has more than 70 % of the top 50 and nearly 60 % of the top 100 universities, which is what Rosovsky assumed in 1990.[404] This proportion has not changed over time. In the 2008 ranking the U.S. still had 36 of the 50 top universities on the Shanghai list.[405] The United Kingdom held steady with about 10 % of these universities. Germany has many at the medium-top. France has strengthened its position, partly through administrative reforms, since 2003. The Nordic countries; Denmark, Finland, Norway and Sweden, comes out very well in this comparison, considering that the region has only around 25 million inhabitants. The Netherlands-Belgium also stands out as a top region.

Historical depth can most easily be found in the U.S. The most used ranking around 1960, the Keniston rating from 1957, was based on judgments of peers all over the country. In a discussion about this ranking Bernhard Berelson has included technological institutions not included in the Keniston report. His list is surprisingly similar to the lists today. The top 12 are: Harvard, UC/Berkeley, Columbia, Yale, Michigan, Chicago, Princeton, Wisconsin, Cornell, and Illinois; to these MIT and Cal. Tech. are added.[406] Berelson also made an investigation of the more prestigious journals in the country, but only for the year 1957–58. He established the following top ten ranking: UC/Berkeley, Harvard, Wisconsin, Columbia, Michigan, Cornell, Chicago, Illinois, Minnesota, and Purdue.[407]

404 Rosovsky (1990), p. 29.

405 In the *Times* list the U.S. totally dominated the top ten with seven universities in 2004 and eight in 2008, but among the 50 top the *Times* list shows a lower proportion for the U.S. with about half of the universities.

406 Berelson (1960), p. 124–126, 280. Also further down on the list, for the following ten, the similarities are remarkable over time. See Bok (1986), p. 15 about the small changes over time.

407 Berelson (1960), p. 127. Below the top ten for instance Stanford is number 12, Yale number 13.

Table 4. The distribution of top universities in the world 2003.

Land	*1–50*	*1–101*	*1–200*	*1–300*
USA	36	59	94	125
UK	4	8	18	26
Japan	2	5	9	14
Canada	2	4	7	16
Switzerland	2	3	6	6
Germany	1	5	13	21
The Netherlands	1	3	8	10
Sweden	1	3	5	9
Australia	1	2	7	8
France	0	2	8	12
Italy	0	1	4	11
Israel	0	1	3	4
Belgium	0	1	2	5
Denmark	0	1	2	4
Austria	0	1	1	4
Finland	0	1	1	2
Norway	0	1	1	1
Others	0	0	10	22

Source: Shanghai Jiao Tong University, Institute of Higher Education.

International ratings, especially between different countries, are harder to find, but Clark Kerr made an attempt in the 1970s. He used two weights, Nobel prizes in the natural sciences on the one hand, and social science breakthroughs on the other. The list he produced, ranking from the top, was: the United States, Great Britain, Germany, Russia, France, Switzerland, Sweden, Denmark, Austria, and Italy. Outside the top ten we find Belgium at 13, Japan at 14, and the Netherlands at 15, but the selected weights mean that the further down on the list the more arbitrary is the placing.[408]

The Nobel prizes can be further used as an indicator of change over time. Of the 140 Nobel prize winners in the sciences from 1901–1945, 24 came from the U.S., and between 1946–1993, 173 of the 310 laureates in the sciences and economics. Of further interest is that of the U.S. laureates before 1945, two thirds were born in the U.S. and three fourths of those who were awarded the prize after the war.[409]

408 Kerr (1978), p. 176. Of the total number of Nobel prizes the United States had 46 %, of the social-science-breakthroughs 53 %.

409 Therborn (1995), p. 262–263. That is: 16 of 24 laureates before 1945 and 129 of 173 from 1945–1993 were born in the U.S.

U.S. dominance was well established in the 1970s, and it has been perpetuated since.

The University of Illinois (UIUC)

Of specific interest for my investigation is the ranking of the University of Illinois, with its main campus at Urbana-Champaign. During recent years it has been ranked around 35–40 in the country on the *U.S. News* list.[410] On the Shanghai list from 2003 the University of Illinois was 35th in the U.S. (and 45th in the world), but since 2004 it has held position 20 in the U.S (26 in the world).

Around 1960 the University of Illinois was among the 12 top universities in the country, according to the estimation by Berelson (and Keniston) referred to above. The main reason for this top placement in the peer ranking was certainly that this university in the mid 1950s was one of only a half dozen universities in the country that were spending more than $10 million on organized research (the others were MIT, Chicago, Berkeley, Michigan, and Columbia).[411] Roger Geiger has used Illinois as an example to illustrate the perils that universities faced in the 1970s. The board started to cut costs at the university in the early 1970s, and the governor even made this a political issue in 1971.[412] Around 1980 the university was accounted among the top 20.[413] The University of Illinois had slipped down a little on the ladder. This was also a part of a larger shift in the university research system. Advancing universities were predominantly those in the South and in the West.[414]

When Dovring arrived in the 1960 it was nevertheless in the top tier.

410 2000: 34, 2003: 38, 2004: 40; 2009: 40.

411 Geiger (1991), p. 207; Geiger (1993), p. 209.

412 Geiger (1993), p. 265. In his historical sketch the university archivist Maynard Brichford talks about "intermittent budget crisis resulted from declining levels of federal and state support for higher education", Brichford (1995), p. 3.

413 Geiger (1991), p. 212.

414 Geiger (1993), p. 268.

APPENDIX 3

Dovring's impact

This biography has focused on Dovring's impact. I have even characterized it as mainly an "influence on", type of biography. I am interested in Dovring's influence to his heirs. I suspect this is the type most subjects would like to read about themselves. Which imprint did the person leave in the world? Academics seem especially interested in this aspect of their own lives.

In academia competition is fierce and different yardsticks are used to measure importance. I have already touched on this subject in the discussion of universities, but now I am turning from the macro level to the micro level of individuals. A myriad of professors creates a university, and to understand how a professor is measured is to reach a deeper understanding of university rankings.

The most common yardstick at present is citation indexes. But a citation index has many drawbacks, especially when it comes to the humanities and social sciences. I will explore a number of indicators, but merely touch on the scientific literature on measuring impact, and then only when it has some direct relevance to Dovring's publications.[415]

Publishing as performance

In the main text I include graphs showing the total number of pages and the number of pages on different subjects that Dovring published. Such statistics could give a frame of understanding for his work capacity and the focus of his endeavors.

The total number of published pages does not reveal much about the spread of Dovring's ideas; instead his influence can be gauged by the venues in which he published. Some of his articles appeared in prestigious journals. Around 1950 the highlight was his article on methods in the French journal *Annales. Économies – Sociétés – Civilisations,* which probably is the best-known history journal in the world. Later, in the 1960s and early 1970s, most of his articles were in the field of economic history, in journals such as *Comparative Studies in Society and History* and *Journal of Economic History.*

415 Pierre Bourdieu, in his famous book *Homo Academicus* from 1984 (in English 1988), uses a number of different indications to measure status among French scholars, such as translations, membership in academies, etc – but not the *Citation Index*, which he considers biased in favor of the sciences. A critical discussion of different American yardsticks used for American academics is Glassick, Huber & Maeroff (1997).

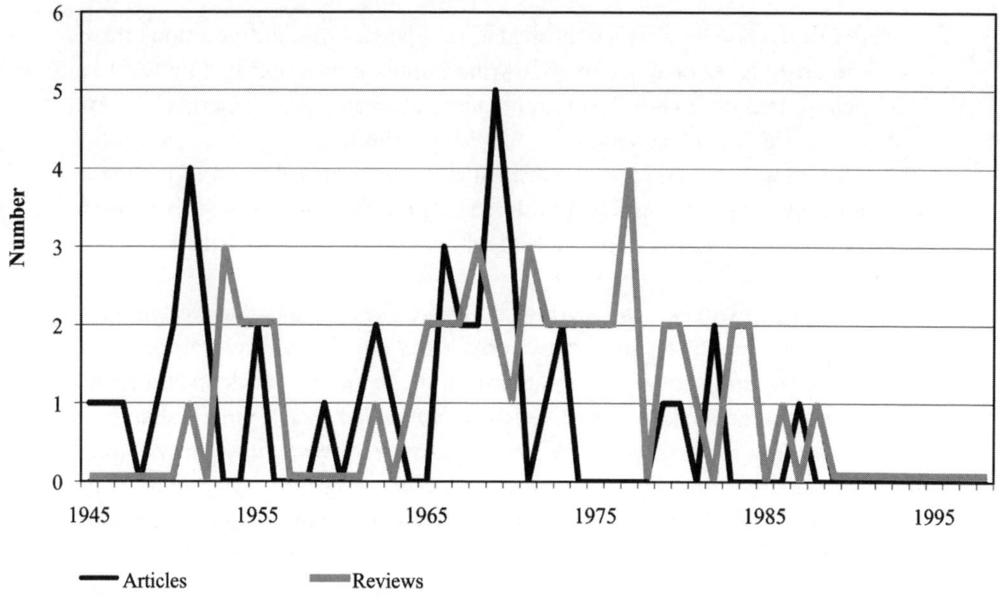

Figure 6. Dovring's articles and reviews published in scientific and scholarly journals.

In an article in the *Journal of Economic History* (1975) Albert Niemi tried to estimate the performance of different universities in the field of economic history by measuring the number of articles published in four leading journals. Niemi explains that journals are "regarded as the standard outlet for disseminating new research findings", but admits that those who concentrate their efforts on books are underrepresented.[416]

The University of Illinois was ranked around number 5–6 among American universities during the period 1960–1974. (At this time UIUC was among the top ten in general rankings.) Of the nearly 50 articles published by scholars from this university Dovring had only contributed two. Among the journals in the study Dovring published only in *Journal of Economic History*, and then mostly reviews, which were not counted.

Another yardstick to measure the impact of books is new editions and translations. The reissue of a book could of course be the result of a very limited first printing, but generally it is a sign of considerable interest, and translations are even more so. Five of Dovring's books were re-published or translated. The most important was *Land and Labor in Europe*, which came out in 1956 and was published in new editions in 1959 and 1965. His book about methods in history from 1960 appeared in a Japanese translation in 1972, and was reissued in English in 1984.

416 Niemi (1975), p. 635. He also counts the number of pages, and furthermore makes a weighted presentation where articles in *Journal of Economic History* are given the double weight.

Half a dozen articles were republished or translated. Outstanding among these was "The Share of Agriculture in a Growing Population", published by FAO in 1959. It appeared in English, Spanish, French, Italian and later in German. It was reprinted in English in 1964 and in Spanish in 1968, and again in English in 1978. (It also was quoted in textbooks, which I have touched upon in the main text above.) Among other articles translated we find, for instance, one about land reform in Mexico published in English and Spanish in 1970, in Italian in 1971 and republished in Spanish in 1974.

The frequency with which Dovring's books were reviewed is a further measure of impact. Dovring's most important book, *Land and Labor*, was reviewed in a first wave of about 25 reviews after the first publication, most of them enthusiastic and many in prestigious journals such as *Economie rurale* (France), *American Sociological Review*, *Agricultural History* (U.S.A.), *Zeitschrift für Agrargeschichte und Agrarsoziologie* (Germany), *Geographical Journal* (England), *Agriculture* (England), *Journal of Agricultural Economics* (England) and astonishingly enough also in the famous French daily *Le Monde*. In the mid 1960s the third, enlarged edition was reviewed in about ten international journals, many of which had already reviewed the first edition.

His book about methods, *History as a Social Science*, which came out in 1960, was reviewed in half a dozen international journals, among them well-known ones such as *International Review of Social History*, *American Historical Review* and *English Historical Review*. After the late 1980s he published a number of books that were reviewed in one or two journals each, not always favorably.

To summarize: around 1960 Dovring had quite a breakthrough with *Land and Labor*; in the late 1960s and early 1970s he published in influential journals.

Citations as a measure

Citing books and articles establishes the sources of facts and figures and allows scholars to acknowledge those who inspired their ideas (or to point to those whose ideas they oppose, which also is a kind of inspiration). A citation index thus seems to be a very good indication of influence.

A world-wide citation index has been built up during the last decades, ISI or Web of science, including *Science Citation Index Expanded*, *Social Science Citation Index*, and *Art & Humanities Citation Index*. Of about 8,000 journals covered in 2003 only one third are from the social sciences and the humanities; the rest focus on the natural sciences.[417] In figure 7 below I have used the citation index beginning in 1982; prior to that the social sciences are largely absent from the index.

417 5300 journals in the sciences, 1700 journals in the social sciences, 1100 journals in the arts and humanities. Before circa 1980 the sciences dominated totally.

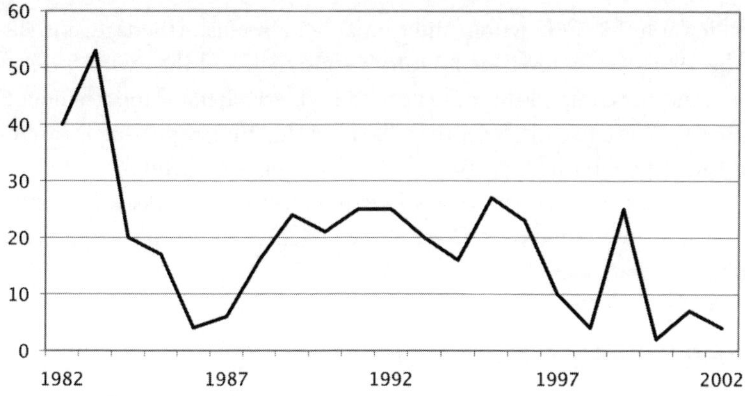

Figure 7. The number of citations per year, Dovring's publications, according to ISI.

There are some drawbacks. Natural scientists often publish shorter articles, whereas scholars in the humanities prefer to present their results in books and will thus be even more underrepresented than the smaller number of journals indicated. Citations in scientific journals mainly measure outreach to other scientists. To get an indication of a more general impact, citations in textbooks would have been a good measure, but then a method must be developed to select the most influential textbooks in different subjects and countries.

Two different measures can be used. One is the number of hits, the other the number of journals. I have used the first mentioned, but also compared with the second. Dovring's number of citations averaged 30–40 per year in the early 1980s, and then decreased to 15–20 for many years. The late 1990s saw a further decrease. Most quoted among Dovring's publications is *Land and Labor*, representing about 40 % of the total. This is respectable but not overwhelming, and when citation of *Land and Labor* decreases no other publication fills the gap.

The number of journals quoting *Land and Labor* is a little more than half the number of citations; that is, the book is on average quoted twice in each source. Dovring's other publications are usually quoted just once per article. The second most cited is his article in *Cambridge Economic History*, written as a complement to *Land and Labor*. The textbook *Land Economics* in its various mimeographed editions is quoted regularly, but the printed book was not much quoted. Among the series of books Dovring published late in life the most often cited was *Inequality. The Political Economy of Income Distribution* from 1991. Other publications are cited only a few times, or not at all.

Dovring might have been interested in a comparison with his brother, the entomologist. Frej Ossiannilsson usually had as many hits as Dovring in the 1980s and slightly more in the 1990s, though his entries also decreased over time. As a natural scientist, however, he was more likely to be cited. Another comparison of interest is with Dovring's old adversary in Sweden, the historian Erik Lönnroth.

He had fewer hits than Dovring, and primarily in Swedish-language journals. Not until the 1990s are citations of Lönnroth registered, and then with 5–10 entries per year. Another comparison could be with Eli Heckscher, the most famous Swedish historian of the twentieth century. He received hundreds of entries even half a century after he published his main works. Dovring is far behind.

In the 1980s and into the 1990s Dovring's historical works were still quoted regularly, also in textbooks.[418] Dovring belonged to the corpus of international agricultural historians.

Reference books and the Web

Dovring has not been included in major encyclopedias, but was mentioned in reference works about important living people. He himself considered such books important and had quite a collection of them. As a measure of impact different *Who's Who* is nevertheless of dubious value, since the business concept often is to sell the publication to persons mentioned in the book. The more inclusive these volumes are, the less interest they have as a measure of impact.

In the Swedish equivalent to *Who's Who*, the more official *Vem är det,* Dovring is listed beginning in 1961, the precipitating factor being his appointment as professor at a major American university. The first international biographical handbook in which his name occurs is *World's Who's Who in Science from Antiquity to the Present* (1968). Among biographical handbooks he next appears in *Who's Who in Finance and Industry* (1992–1993). The immediate impetus seems to be his campaign for farming for fuel in a number of radio programs for Voice of America in 1990. In *Who's Who in America* Dovring is included in 1993–94, but as there were more than 80,000 entries in this edition and entries vary from year to year this is not a good measure of impact. More selective is *Who's Who in the World,* with about 20,000 to 25,000 individuals selected for lifetime inclusion. Dovring was added in 1993–94.

With regard to these international biographical handbooks, Dovring surpassed his friends and antagonists at home in Sweden; his friend Torsten Husén and his older brother Frej Ossiannilsson are both mentioned in *Who's Who in Science* (1968), but not in any *Who's Who in the World.*

A scholar and author like her husband, Karin Dovring was affiliated with Yale University from 1953–1978. She also published novels and short stories (in English). As early as 1977–1978 an article on her appears in *Who's Who in American Women.* She is mentioned in *Who's Who in America* 1992–1993, one year before Folke. Karin Dovring was included in the selective *Who's Who in the World* in 1984–1985, long before her husband.

---

418 See above chapter 4, and for instance also Jones (1988), p. 71 about peasants as rational cultivators, quoting a review Dovring wrote in *Comparative studies in Society and History* 1962.

It is an irony that this kind of fame, mention in biographical handbooks, of which Folke Dovring was so proud, primarily came to his wife. Perhaps it is a sign of the times that intelligent and able women in culture and academia were at last being acknowledged.

An even more unreliable measure of impact is the Web. Checks during 2002 using the search engine Google gave more than 600 hits. Most of these were related to Dovring's works in the field of economics and history, but a considerable subset of the highest ranked hits concerned his discussion of cannabis in the book *Farming for Fuel.* These quotations are often located in home pages advocating free pot, which Dovring had not promoted. The Web is sometimes a distorted mirror of the real world. A new check using Google in late 2004 gave more than 800 hits, but now cannabis had fallen into the background and historical articles held a rightful first rank. And a last check in 2009 gave him over 2000 hits. The Web is expanding, and in 2009 most of the hits in the top were related to his books as they appear in lists from antiquarian booksellers. I must also admit that many of the hits in the top of the ranking were related to my research project and different articles I have written.

Some concluding remarks

One could well argue that numbers do not say much. The hypothesis I wanted to test was if Dovring's later books were forgotten. Most of the statistics mentioned above appear to sustain such a hypothesis, but antiquarian book sellers nevertheless seem to consider even his later books of such interest that they may attract customers.

In the opening phrase of this book I refer to the Icelandic sagas – and I have named the whole book after them: *The Dovring Saga*. In Havamál there is a famous saying: "Men die, creatures die, but I know one thing that does not die: the reputation of a dead man." This threat haunts us all, as it was meant to do. But is it correct? The deeds we have done are there even if forgotten, and the aim of biographers and historians is to see to it that the reputation of the dead does not die – which it otherwise will.

APPENDIX 4
Dovring as social reformer and economist

By Gabriel Söderberg

In 1960 when Dovring settled in America, fifteen years had passed since the end of World War II. Europe had emerged from the ruins of countless air raids to find itself divided by what Churchill famously called The Iron curtain.

The Cold War was never just about the rivalry between two superpowers in the post-war era. It was the rivalry between two economic systems, two world-views, two bids to shape the future of mankind. Living in America was not just a geographical denomination: it was a way of life, it meant being part of one of the two economic systems that the rest of the world used as blueprints to plan their future. American society was a model to be emulated. Not just American exports, investments and subsidies poured into Western Europe after the war ended; they were followed by a flood of American movies, music and consumer products. Living in America was being part of an economic system as well as a cultural hegemony.

The Cold War was thus largely a question about which system was most efficient. Which system could provide the masses with the best entertainment? Which system would be the first to send a manned expedition to the moon? Which system could build the most massive arsenal of nuclear warheads?

Living in America was thus to choose sides. It is certain that Dovring chose to live in America not just because of opportunity or necessity, but also out of ideological concerns and a strict view that the market-based economy of the West was superior to the centrally-planned Soviet system. He condemned the Socialist experiment both as fundamentally repressive of freedom and as appalling in its economic inefficiency. The American way of organizing society was superior to its rival; this was beyond doubt in Dovring's mind. This did not, however, mean that the American way of life and the American society was good enough. Being better than the Soviet system was nothing to be proud of. "The American system", he wrote, "is without doubt far better, but then it should be. Soviet socialism has created such a mess of built-in inefficiencies that we should not draw too much comfort merely from being better than that."[419]

Here he was then, Folke Dovring, aged 44 – not the age to settle down and rest on one's laurels – starting a new life in the country whose system was the number one contender to shape the future of mankind and was still not good

419 Dovring (1984), p. 147.

enough. While still a young man in Sweden, he had proved himself to be a trouble-maker, not afraid of making enemies, and perhaps not the most diplomatic person you could find. He was hardly afraid of making a controversial figure of himself. Being a European by birth he could observe the American society with the eyes of an outsider, thus placing some distance between himself and the object of study which benefited a clear-sighted analysis. With time he became increasingly convinced that there were several under-currents of sickness deep at the heart of the American society, the worst of this being a systematic tendency to wastefulness that threatened the sustainability of the American way of life itself. This had to be corrected. The last decades of Dovring's life were dominated by his critique of systematic errors in the American system and hope for its reform.

But how can we understand such an ambition? When I began reading Dovring's writings I was confused. Here was a man who in his youth had made it his business to study a subject that most of his peers denounced as lacking in dynamics, utterly void of progress and thus without history – namely rural society in premodern times. He had grown up in the countryside and brought with him a traditional morality that made him recoil at the wasteful use of resources in modern American society. Yet, I realized, Dovring was a firm believer in modernity, science and the progress of mankind through the continual growth of wealth. He had a strong belief in the human mind and its capacity to keep pushing the boundaries of what is possible further and further. He adored technology, and was fond of speculating about what breakthroughs and discoveries might lie just around the corner or hundreds of years into the future.[420] Yet here he was, a firm believer in modernity and progress, condemning much of what many of us consider the very essence of modernity and progress – our consumption habits, our freedom of mobility through cars and our lack of worry over scarce resources. Was this not a contradiction, the ramblings of a confused personality without a coherent world view?

In the pages to come, in which Dovring's thoughts on reform and economics are explored, I shall argue that a coherent web of thought actually does emerge. In his critique of society Dovring challenges our very conception of modernity and progress. What might be termed a traditional fixation on the need to conserve scarce resources is not just about ensuring the survival of the modern project. The key to progress must be viewed as the constant improvement of efficiency and the gradual process of creating more and more out of less material input. Thus a wasteful society, in Dovring's mind, is actually a primitive one and many of the things we usually consider the hallmarks of modernity – the automobile, the supermarkets crammed with goods and the huge amounts of meat in our diets – are actually signs of degradation of progress. Progress, instead, entails a constantly leaner basis of material input, as technology and design are constantly refined. In that way, the traditional morals of the peasants of preindustrial society

420 See for instance Dovring (1987), p. 191.

are actually an important guideline not only in avoiding destruction but also in order to create the most modern society possible.

But Dovring's thoughts and critique did not exist in pure isolation. The decades at hand – 60s, 70s, 80s – were decades of turbulence and change. Dovring was not the only would-be reformer and far from alone in taking a radical stand against mainstream society. The youth movements of the 1960s, the hippies and the anti-war demonstrators, the Black Panthers and the Nation of Islam, Greenpeace and the ecological communities of flower children – all these were currents that shaped American life which served as the background to the 1980s of "Reaganomics", Wall Street fever and increased economic inequality. These were the decades of Dovring's reform project and must be seen as the stage upon which he tried to convey his message in a cacophony of voices that threatened to – and finally did – drown him out. Thus no presentation of Dovring's thought would be complete without placing him in the correct context and as we proceed we must keep these circumstances in mind.

The Optional Society

Reformers often have merely a hazy vision of what society should be like. It is easier to lash out at particular errors in the surrounding environment than to construct a conceptual basis for the good society and judge the extant social structure on how well it coincides with it. In 1971 Dovring, together with his wife Karin, published such a conceptual basis. The book was called *The Optional Society* and contained not only a presentation of the couple's ideal society but also their view on world history and human nature. The theoretical aspects of the work are Folke's, while Karin grounded it in everyday experience by seasoning his scholarly text with anecdotes from the real world.

What, then, was this Optional Society of the Dovrings? As the name suggests it is a society that allows the individual many options in deciding what type of life to lead. "The key word of it all", write the Dovrings, "is choice …"[421] Indeed, the traditional society that Folke had studied as a medievalist had been utterly void of choice and option. One's place in society was given at birth and the type of life that ensued was as fixed as the stars in the sky. "For the common man, lifelong poverty gave little choice but to worship the gods, follow the mores, and cultivate the arts which their community would approve of and could afford; and of course obey its laws and institutions as if they were God-given inevitabilities."[422] Modern society stood in stark contrast to this. The rise of the modern world had liberated parts of mankind by shattering the chains of material need. Marxist historians were wrong in their belief that human culture and intellectual pursuits arose from a material substratum. These were in fact just as basic to human nature as the

421 Dovring – Dovring (1971), p. 6.

422 Dovring – Dovring (1971), p. 3.

need for food and shelter, but lack of material necessities had constantly kept human creativity, the natural pursuit of culture and intellectual curiosity in check. Increased affluence, then, served to remove the obstacle of material need and thus to create increased freedom of choice between more options. The rise of mankind was therefore the increase in wealth that allowed human culture and intellect to work unhindered by material constraints. With this followed an increase in the number of options available for choice. "The expansion of choice", states Dovring, "comes from the material affluence and grows with it."[423]

Modern society thus outshone its traditional predecessors: "Not only is there more to choose from: there are also many more people who have (or could have) a choice."[424] This is indeed the optimist and modernist Dovring, the Dovring that rose up in revolt against his father and patriarch, confident that the ingenuity of the human mind could conquer whatever hindered its true capacity. He saw the history of human achievements as a heroic continuity that was crowned with success in the modern achievement of industrialization: "The intellectual and technical sources on which modern affluence draws have been long in the making. They have been under way since the conquest of fire."[425] And with the advent of industrial production, finally the Malthusian specter that had haunted mankind since time immemorial was exorcised: "The threshold effect of modern industrialism came about because this industrialism made it possible, as it has never been before, to make production of life's necessities grow faster than the human population."[426]

But the process did not end there. "The fact is", Dovring wrote, "that we are heading for – and to some handsome extent already are in – an optional society, one where people may come a long way toward choosing the kind of life they will live."[427] Within increased productivity lay the key to increased freedom: "Modern economic progress may at last turn out a society where mass production of essential goods requires only a small part of all our time and effort. Then we have the option to devote the rest to our individual interests and cultivate our personalities in the degree they can stand a spiritual civilization. A utopian dream? It is closer than you think."[428]

This, then, was the optional society. It was thought to be the best possible society of any individual, because it was the society that best suited the nature that the Dovrings ascribed mankind. For the first time society was approaching what ought to be the natural habitat for a human being. "The key word of it all is choice, and the human mental apparatus was made for choice ...The neurons of

423 Dovring – Dovring (1971), p. 2.
424 Dovring – Dovring (1971), p. 1.
425 Dovring – Dovring (1971), p. 3.
426 Dovring – Dovring (1971), p. 4.
427 Dovring – Dovring (1971), p. 1.
428 Dovring – Dovring (1971), p. 16.

our brain, numbering some one hundred thousand million, make it possible for us to handle unspeakable amounts of information and to reach astronomic numbers of variant conclusions, attitudes, and combinations of data. For the first time we may be able to use in full this brain capacity which has existed a million years."[429] But this did not mean that the process was inevitable. There were potential clashes along the way to attaining full realization of the optional society. Dovring was to see several currents that had been institutionalized into the system of American society that actually eliminated options and thus worked against the realization of the optional society. A large part of his reform project must thus be seen as attacking these institutionalized threats to the optional society.

But there was also a more natural hindrance to the full utilization of this unprecedented freedom of human ability. The plethora of choices that opened up to modern man also meant greater uncertainty. Faced with choice many people felt threatened and in the worst case even wanted others to choose for them. Choice was given to them by the workings of economic progress – it was after all nothing they had chosen for themselves. "The question", Dovring mused, "is why this should have to mean discomfort. Human nature was made for choice."[430] Dovring did not, however, enter the realm of existential philosophy but settled for an answer close to historic materialism. The anxiety people felt when faced with freedom was not the anxiety of Kierkegaard or Heidegger; rather it was "a hangover from past ages of penury. In those times, change would often threaten the whole material existence of many individuals."[431]

Here then was the mechanism at work behind many of the disturbing trends of American society. The hippies that dropped out of society, the drug-addicts that entered a numbing haze and the monotony of modern pop culture – all these were the results of human frailty that could not deal with freedom and chose a blindfolded existence instead. Dovring's later years would involve much frustration because he felt that people could not use their freedom as they should. If hippie collectives and drugs claimed only a small portion of America's people, a monotonous way of life with pop music, hamburgers and mindless entertainment claimed a huge part of the rest. It was the classical dilemma of the reformer – reform must be initiated so that freedom and opportunity can flourish, but the common man keeps insisting on using his newly-won freedom and opportunity to roll around in the dirt.

The monotonous lifestyle of mass consumption, which blocked many of the options that opened up to the modern individual, went hand in hand with another threat to the optional society. This was the wasteful use of resources in America.

---

429 Dovring – Dovring (1971), p. 6.

430 Dovring – Dovring (1971), p. 6.

431 Dovring – Dovring (1971), p. 6.

Waste

Late in life Dovring had occasion to reflect upon the experiences that had shaped him. Many of his interests and concerns went back to his childhood "in the middle of an old-fashioned peasant country."[432] One could think that he left all this behind him when he revolted against his father and the patriarchal traditions of his time, but that would be to underestimate the complex nature of the human mind. Freedom of the individual against all sorts of conformism – including traditional ones – was a cornerstone of Dovring's thinking to the end. But that did not mean that he turned his back on all of the values and attitudes that surrounded him as a child.

This is evident in Dovring's feeling about the concept of waste. Here he is firmly rooted in the morality filtered through generations of farming communities. It goes back to the original meaning of economics, which is the art of keeping a household in balance. Sooner or later both ends must meet. You reap what you sow, you preserve what you do not need for the time being and – most important of all – you *do not* waste. It was the morality of the subsistence economy which he had imbibed as a child – waste was wrong in itself, both immoral and dangerous for the survival of society. This stands in contradiction to what many of us consider the logic of modern society, in which increased abundance promises to deliver people from the need to worry about the scarcity of resources.

Throughout his life he had a tendency to view the *oikos* in the ancient Greek sense, that is as a self-subsistent economic unit, as the ideal for a modern national economy. In this he might have been influenced by Swedish geopolitician Rudolf Kjellén, whose concept of autarky – that is complete independence of resources outside of the nation's borders – was very influential in European strategic thinking when Dovring was a young man. Dovring referred to such an economy as "a complete economic system". The British Empire had been such a complete economic system, thanks to its colonial possessions. America – with its vast expanse of land and its huge endowments of natural resources – had also been a complete economic system until the 1950s when increase in imports led to dependence on resources from other parts of the world.[433] In this – as we shall soon see – the frivolous use of oil to feed the growing use of cars in America was pivotal, but also a wasteful attitude inherent in the American way of life. The morality on which this way of living was based stood in stark contrast to the traditional self-subsistence morality of Dovring. It was the morality and the way of dealing with resources of a "frontier society", of a society that is constantly discovering new unclaimed riches of land and resources, constantly moving on, constantly expanding. The American history of settling the continent coast to coast – the Manifest Destiny – had ingrained the frontier mentality deep in the heart

432 Dovring (1998), p. vii.

433 1975, unpublished.

of the Americans. It had made them a people prone to waste, and this waste made them vulnerable and dependent on resources from the outside world.

But waste was also a highly efficient strategy for producers and owners of resources to keep up the prices of their goods. "Social waste", Dovring wrote in *Riches to Rags*, a sort of summing up of his ideas on waste published in 1984, "is used consciously, and on a large scale, to sustain prices which tend to become less scarce and less expensive. Such waste hurts society as a whole and most of its members, but it favors those who own the resources that are wasted."[434]

Here then was a form of economic inefficiency that mainstream economics had failed to take into account. In mainstream economics there is the concept of monopolistic practices that serve to maintain high prices by reducing production. Dovring claimed that there was yet another form of price-manipulating behavior that harmed the total welfare of society. This behavior worked to hinder the natural tendency of economic progress – mostly through advances in technology – to achieve a decline in the minimum requirement of any given resource. This might mean a sharp decline in prices for its use, something that the owners would want to avoid. "This gives the cue", Dovring wrote, "to resource owners as to how they can create redundancies and so avoid the consequences of declining costs and prices. Without counter moves by resource owners, many of them would cease to be rich, even as the general public would be better off. Owners of anything that is scarce must therefore wish it to continue to be scarce."[435]

Instead of manipulating the supply side of the economy – as is done in restriction of production – the resource owners manipulated the demand side. They encouraged excessive use of their resources and made sure that the market for threatening substitutes was kept closed. Waste achieved the goal of keeping prices high by making sure that resources were not used with the greatest possible efficiency so that demand was constantly higher than it would be if efficient use was actually made of them. "The result of such manipulation of consumer (and investor) demand is social waste."[436]

The frontier mentality of people in America – where the vast resources of the continent "encouraged, early on, waste rather than thrifty use"[437] – went hand in hand and the two taken together painted a dark picture of the future.

But if waste was such a big problem, why then was it not obvious to more people, why was there no political dispute about how to deal with it? The answer was that waste was built into the system itself and therefore achieved a false aura of rationality. Indeed, the logic of Keynesian economic thinking suggested stimulating demand in order to keep employment up – even if this meant using resources in a way that seemed wasteful to common sense. Keynes himself wrote in

434 Dovring (1984), p. 1.

435 Dovring (1984), p. 2.

436 Dovring (1984), p. 2.

437 Dovring (1984), p. 3.

his *General Theory*: "If the Treasury were to fill old bottles with banknotes, bury them at suitable depths in disused coalmines which are then filled up to the surface with town rubbish, and leave it to private enterprise on well-tried principles of *laissez-faire* to dig the notes up again ... there need be no more unemployment and, with the help of the repercussions, the real income of the community, and its capital wealth also, would probably become a good deal greater than it actually is."[438]

Such economic thinking – that waste stimulates the economy – was anathema to Dovring. It could be advantageous only in the short run. To refute it Dovring used a similar argument to that used by proponents of free trade when faced with the pressure of foreign competition: "Whenever we try to imagine some kind of waste cut down, we not only would reduce some current employment. We would also set free some resources which can be used to increase our total satisfaction and to increase the employment in producing those other satisfactions. Eating less wastefully would save land for a healthier landscape and for benign energy sources. We would also save fuel and chemicals for other uses and for a less polluted environment and we would save labor for work on something else. Using transportation less wastefully would free hardware, fuel and manpower to do things we now 'can not afford'..."[439]

But Keynesian economic policy, a frontier mentality and manipulation by vested interest were not the only roots of the problem. There was also a tendency in all economic systems to be highly dependent in their daily functions on decisions taken early on that set the course of things and then bound them to the same patterns. That meant that one could not expect the market system to allocate resources entirely effectively, since all things that happened in the market were constrained by an institutional framework that had, thanks to past decisions, established wasteful procedures. In this he was a staunch believer in the so called institutional school of economics – originally an American school with the economist Thorstein Veblen as one of its founders. The attacks upon mainstream economics that Dovring made from this position will be dealt with in detail later. For now it is enough to know that Dovring thought that the American system – or set of institutions – was rigged in a highly wasteful way that left little room for market or individual behavior to solve the problem. Indeed this was probably the most important part of Dovring's theory of waste: "Our main theme in this book", he wrote in *Riches to Rags*, "is to show how waste is enforced by built-in constraints which often leave the individual little or no choice."[440]

He did not belittle the role of the consumer in shopping in a more enlightened way to reduce waste. But there was only a small part of the waste that could be dealt with through consumers – the bulk of it all came from the workings of

438 Dovring (1984), p. 129.

439 Dovring (1984), p. 5.

440 Dovring (1984), p. 8.

the production systems that were trapped in a prison of procedures. These "waste-enforcing mechanisms" formed "coherent webs, 'multiple lock-in systems', which are not easily broken by individuals or individual firms, or even by communities acting in isolation from other communities."[441]

Already in *The Optional Society* Dovring had described how an economy can build a prison for itself through these lock-in systems that threatened the good society. "By its very nature", he wrote, "such a system tends to perpetuate itself even when its original reasons are no longer valid."[442]

The American economy functioned in ways that had been laid down in the 1950s – a decade that many, including Dovring, identify as the decade in which things went wrong – under circumstances wholly different from those of the 1980s when Dovring wrote *Riches to Rags*. The wastefulness of the food industry, military expenditure, the school system and the legal procedures were all shaped by this. But the worst case – and the most typical example of a lock-in system that eliminated alternatives to itself and thus kept the economy in a firm grip – was the oil industry that together with the automobile presented a nearly impregnable fortress of unsustainable wastefulness. It is so important in Dovring's story that it deserves a section of its own.

Dovring's attack on the oil/car complex

Compared to the self-sufficient economy of the *oikos* or the traditional farming community, the American way of life with its heavy reliance on huge amounts of oil was a disaster. Not only did it result in vulnerability to supply shocks that sent inflation soaring domestically but domestic policies were to enormously dependent upon the price of oil, which restricted the freedom to act politically. It also meant that American foreign policy was perverted from what it ought to be – a stabilizing force – into strategic meddling in oil-rich regions to keep the black gold flowing. "The entire web of Middle East politics", Dovring wrote, "depends on this one factor. Without this unnecessary import demand, the United States would have had a much freer hand to pursue its foreign policy according to essential critera of international order."[443] The freedom to choose the nation's course was thus diminished by its dependence on oil, something that was not in accord with the increase in options that increased wealth should bring.

That was bad enough. But the real problem was that the oil was running out. The central place occupied by a finite resource about to enter a new phase of scarcity threatened the entire system. "The belief that fuel will always be available at some price seems to disregard the fact that there can be fuel prices that can

441 Dovring (1984), p. 8.

442 Dovring (1971), p. 88.

443 Dovring (1984), p. 35.

break the back of any economic system."[444]

Arguments that society could adapt to constantly higher oil prices through market efficiency that provided incentives to solve the energy problem as time went on, were rejected by Dovring on the basis of his belief in institutional economics and the wastefulness that was built into the system and hindered market adjustment to this new future. "Leading circles in Washington", he noted, "continue to insist ... that market mechanisms will solve our economic problems, including that of energy, if they are only left free to operate without political interference. In thus treating the market as if it were a force of nature, these people appear to overlook the fact that any society always operates under some set of institutions which are not necessarily what current market forces would choose if they had an entirely free choice. Rather, these institutions (reflections of past policy decisions) in many ways determine how current markets may function."[445]

Indeed, the fateful love affair between America and oil went back in time. Whales had been harpooned off the coast of Eastern America – described in the novel *Moby Dick* that is a corner stone of American literature – and oil from these mammals used to fuel lamps. The idea that mineral oil could be used for the same purpose and extracted on a large scale for mass consumption was revolutionary to the course of mankind. The first drilling took place in Canada, but American enterprises soon followed suit. In August 1859 in Titusville, Pennsylvania, the drilling paid off. Oil began to be pumped up from a depth of twenty meters.[446] The riches thus uncovered below the ground were far greater then the fueling of lamps could possibly tap to its full extent. This, to Dovring, was sheer accident. "In the past, America was incredibly lucky. Petroleum, and later natural gas, became available in quantities which initially were far greater than anyone had use for and the costs of extraction were trivial compared to the marginal product generated by putting those energy sources to work."[447]

These highly special circumstances – which would never come again since the chance of finding another great reservoir of oil was quite small – led to decisions being taken that geared the economy into the wasteful institutional pattern described earlier. The sensation of having practically infinite amounts of oil coursed through the budding car society that took shape in the early 1900s and culminated in the 1950s, when a number of key political decisions were taken that ultimately sealed the dangerous pact of oil, car and political interest. Instead of starting to move the country towards more efficient management of resources, the government chose to charge ahead with an economy that assumed the same abundance that met the first oil-drillers in 1859. The results were massive construction

444 Dovring (1984), p. 39.

445 Dovring (1984), p. 38.

446 Lindstedt (2006), p. 67.

447 Dovring (1984), p. 38.

of superhighways, continually increasing numbers of cars and miles driven as driving became an everyday experience for the average American. "The legacy of the 1950s", Dovring wrote, "is cast in concrete in the over-large system of superhighways and the too great dependence on trucks and cars. The 1960s largely continued on the momentum set in motion in the 1950s."

This was a typical example of a multiple lock-in system. The average consumer did not have much choice in this as the possibility of using collective transport declined. "People were gradually pushed off the trains and buses whether they liked it or not. Demand for cars rose more than would have followed from freely expressed preference. Many of the new connections were indeed shotgun weddings."[448]

Thus what could be called economic growth ironically eliminated options for individuals and placed constraints on their way of life – in this case each American *had* to own a car in order to function; this was a pressure to conform that worked against the creation of the optional society. And as a multiple lock-in system, these procedures tended to perpetuate themselves even now when oil was dwindling. This system, built on use of oil as if it was not a scarce resource, was about to meet a new era. "Our most recent experience shows that now the marginal costs for oil are very high and may be rising toward intolerable levels. When current policy merely encourages the oil operators to go on drilling without regard for the cost of the national economy, it is clearly steering the country toward very hard times in the medium-term future."[449]

The only way to solve this was a huge break with the past. But the connection between vested interests and politicians together with the sheer cost of the reorientation made this difficult. Politicians "dare not speak out to voters on the economic changes which are necessary to get us out of the downward suction from high energy costs. When voters hear nothing much about it from their politicians, they feel justified in not asking for changes either. In this way the sociopolitical situation, like the housing-and-traffic system, becomes a "multiple lock-in" system, the breaking of which will necessitate a major reorientation."[450]

Dovring took it upon himself to issue the wake-up call. Many of those who heard about him in this way must have thought he was a very odd person, a doomsday prophet and as such – since the world is still standing despite several prophecies of Armageddon – most likely wrong. "The danger of economic collapse in the United States", he wrote in the *Washington Post* in 1976, "within the next 10 to 20 twenty years is a very real consequence of the energy problem."[451]

Two years later he published an article in the local newspaper *The Morning Courier* entitled *Declare War on Energy Problem* in which he declared that "Amer-

448 Dovring (1984), p. 48.

449 Dovring (1984), p. 38.

450 Dovring (1984), p. 41.

451 Dovring (1976).

ica's energy problem is the economic equivalent of war, because the nation's continued existence is at stake and only extraordinary efforts can carry us through it."[452] Indeed what was at stake was nothing less than the survival of society: "Unless we save energy on a grand scale we may, in a decade or two, see the nation's economic and social fabric go to pieces, never to be put together again as a free society."[453] No amount of wishful thinking or optimistic beliefs that oil reserves would be sufficient could change this. Drastic measures had to be taken. New research and investments in alternative energy on a colossal scale had to be implemented. The resources needed could not be directed to this purpose through market mechanisms any more than a huge military build-up could be achieved by the market. Instead the government had to take control over large parts of the economy. "A partial war economy, with administrative measures to re-direct our economic processes away from the present collision course with disaster, must be started without further delay."

This is extremely radical thinking – to ordinary Illinois readers, sitting at the breakfast table with their morning coffee, it must have bordered on madness.

The main question at this time – some years after the oil crisis of 1973–1974 – was whether American reserves could go on delivering enough oil in the future. To Dovring it was certain that they could not and that the OPEC supply could not be trusted, since these countries themselves would soon embark on oil-guzzling industrialization of their own. Therefore much of the discussion turned on the field of geology rather than pure economics. Intellectually Dovring was – or tried to be – something of a jack-of-all-trades, but in arguments on, for example, the nature on different types of oil and oil layers it seems he was out of his depth. This was pointed out by geologist Donald C. Bond in an attack on Dovring that accused him of being ignorant of many of the details in practical oil drilling which led to faulty conclusions.[454] Dovring – never a person to take criticism lightly – responded by accusing Bond of being too enmeshed in the oil industry to be trusted in these matters.

Was Dovring right in his claims? It is evident of course in the year 2007 that the American economy did not collapse ten or twenty years after Dovring's article in the *Washington Post*. On the other hand the consensus is that American oil production actually did reach its production peak in the early 1970s and that the great confidence of people like Bond in domestic oil reserves was misplaced.[455] What Dovring underestimated though was the power of American foreign policy to keep the oil flowing and the inability of the OPEC countries to industrialize. On the other hand he did see that oil interests changed American foreign policy, so he was without doubt partly right in his analysis of the oil situation.

452 Dovring (1978).

453 Dovring (1978).

454 Dovring (1982).

455 Deffeyes (2001), p. 4.

Nowadays the discussion has extended to a global level. The question is whether the total oil reserves of the world are being depleted – or rather whether the production peak has been reached or soon will be– so that the cost of oil is going to sky-rocket. Calculations that successfully predicted the American peak in 1970, predict that the peak for world oil occurred around the year 2000.[456] If this is true, we will now face constantly higher marginal costs for pumping oil as the pressure in the reservoirs lessens due to less oil underground. Thus the so called "peak oil" discussion is mainly a continuation of the discussion that Dovring was engaged in. The success of American oil policy made sure that the depletion of American oil reserves did not mean the end of the way American society worked. But depletion of the world's total reserves is of course a totally different question and – as Dovring would have said – no amount of wishful thinking makes the problem go away.

Discrimination and racism as waste

Another important aspect of Dovring's thinking is his view that racism and inequality between the sexes was a tremendous waste of human talent that hurt society itself.

When Dovring settled in America in 1960, the country was plagued by racial tensions. Men like Malcolm X and Martin Luther King served as figureheads for the hopes and frustrations of millions of black Americans. Violence was in the air. In 1965 the Watts Riots in Los Angeles set off a number of riots across America. Both Malcolm X and Martin Luther King were assassinated. Events like these were bound to have an impact on a foreigner from a land of homogenous ethnicity settling down in this country of riots and ghettoes.

Dovring did not believe in natural inferiority in either women or blacks and saw discrimination of these groups as a waste of human talent fully comparable to the waste going on in the use of oil. "Discrimination", he wrote, "against women and minorities has long been supported by pseudo-evidence on purported differences in native ability, between the sexes and between the 'races'. Usually, such evidence crumbles when exposed to critical scrutiny: the statistics have been biased by the results of past discriminations."[457]

In *The Optional Society* he concluded that lack of economic opportunity could mean that the talent of a human being could go to waste because it could not be nurtured.[458] But the rise of the affluent society should have removed all such obstacles. Even so, society kept wasting the valuable resources of human ingenuity and talent by denying individuals of certain groups the scope to use their talents to the full benefit of society. This was especially bad since the complexity of the

456 Deffeyes (2001), p. 4–5.
457 Dovring (1984), p. 100.
458 Dovring (1971), p. 2.

economy had increased to the point where knowledge and skill had become scarce resources. In earlier times, when humanity was shackled by poverty and had the knife of necessity at its throat the need for talent was negligible. "Peasant society with its low productive food farming needed to tie nearly all hands to the soil. Latent brain power threatened to get in the way unless it was trained into mindless submission. The need for and the use of high talent was so small that it could without difficulty be filled from the ranks of the privileged."[459] Even in early modern times, when the foundation of economic thinking was laid, the need for talent was too small for anyone to realize its full value. Classical economists had only listed land, labor and capital as production factors and left out management, which had since become increasingly important. "This may be", he mused, "because in those days managerial talent was surplus, hence not a scarce good, and economists deal only in scarce goods." Now things were different and discrimination much more damaging.

The complexity of modern society meant that it needed all the talent it could get. Racism and discrimination were terribly wasteful ways of managing a scarce resource. This wastefulness harmed society for the benefit of an extremely small group. "When a mediocre white male gets a position that a brilliant woman or minority man would handle better, we are all shortchanged, for we are less well served in this way. Even those who reap but modest positive rewards in this manner may in fact be more hurt by the damage done to the system than favored in their individual lives ... Only a few individuals are really better off under discrimination, and these individuals are then in some sense social parasites."[460]

Reform?

The picture of the good society, then, was clear to Dovring. It was the Optional Society. Also, he had a clear picture of the trends in American society that threatened the realization of this good society. How then was the system in its present state to be reformed?

The major reorientation of the entire economic structure to deal with the energy problem has already been discussed. It could not be stressed enough that the whole system was rigged in a wasteful way: "Rather than blaming anyone in particular, we should look at the system – we must see how it works as a system. Individual firms, even the large ones, only do what comes naturally in their business situation."[461]

What was needed then was "national economic planning".[462] Dovring was careful to point out that this did not mean a move toward a socialist organization

459 Dovring (1984), p. 101.

460 Dovring (1984), p. 21.

461 Dovring (1984), p. 157.

462 Dovring (1984), p. 157.

of society. Indeed he had concluded early in his life that this sort of organization was hideously ineffective. "Rather than discussing the concepts of nineteenth century Socialist thinking, which have long proved their practical failure, we might contemplate the notion of 'guided market economy' which was the watchword in the economic reconstruction of West Germany after World War II."[463]

The key to this guided market economy was to realize that government and business had specific functions to fill and that they should be kept distinct from each other. "What American private business does well, is to run the businesses; what it does not do at all well is to run the country. The latter requires unified economic policy objectives and a balancing of conflicting interests."[464]

Such a guiding set of government regulations would replace the short-sighted incoherent regulations of the present economy with a consistent national economic policy. This would be beneficial for private enterprises as well since they would face "less uncertainty than with today's many conflicting piecemeal regulations."[465] But in the end the main purpose for such economic policy must be the solution of the problem of waste. This was the only way society could go on evolving. "The removal or at least the sharp reduction of social waste will be beneficial in many ways, but it is first of all needed to set the economy on a long-term path of future growth which need not be interrupted by man-made crises."[466]

That stated, however, Dovring seems to have been less certain about exactly how the guiding principles would solve the problems. In *Riches to Rags* there are a number of suggestions. The excessive use of cars must be attacked through a strategic blow to its center, namely the availability of parking space. "The central lever which will control both traffic, fuel use, and city layout is the parking system ... this kind of control is low cost, easy to enforce, and highly efficient."[467]

Further, equal rights and more equality of opportunity must be given to all citizens of the country to reduce the waste of human talent. This also meant that poverty in the Unites States had to be combated, so that the resources of the ghettoes could be tapped into – to the gain of both individuals and society itself. Also the education and legal system had to be reformed and the food industry made to work using entirely different procedures. In this the soy bean was of great importance. The raising of cattle for meat production was wasteful and protein from soy beans should replace that from meat in people's daily diet. In this he earned some much-wanted attention, being the featured writer in an issue of *Scientific American* – a prestigious scientific magazine – in 1974.

His most detailed description, though, of a concrete reform was his plan to

463 Dovring (1984), p. 157.
464 Dovring (1984), p. 158.
465 Dovring (1984), p. 158.
466 Dovring (1984), p. 158.
467 Dovring (1984), p. 159.

use the soil and labor of the American farmers to produce huge amounts of biomass out of which methanol could be extracted. This methanol could then replace oil as the main fuel of the country. The brilliance of the plan, to Dovring, was that several of America's problems could be solved by implementing it – petroleum imports, farm surplus production, soil conservation and air pollution. "What have these four problems in common?", he wrote in the book *Farming for Fuel* published in 1988. "They all can be solved or greatly reduced by one single large policy departure: large-scale production of methanol (wood alcohol) from biomass, to become a mainstay of the energy system of the United States."[468]

Most certainly he saw these practical suggestions as just a few guidelines and expected others – once won over to the great cause of reform – to map out more detailed routes. In the end the goal had to be a self-sufficient economy, towards which one major step was a drastic reduction of waste. Sooner or later, though, only the taming of the only truly stable energy source could solve the problem: "The eventual long-range solution of the energy problem will have to be in some self-renewing system of solar energy conversion which alone can become stable over an indefinite future."[469]

Basically, he was an optimist to the end. He trusted human ingenuity to be able to solve all problems if only given the opportunity to do so without hindrance from artificial or material barriers. Given the right policies, "the economy may very well extricate itself from all its current difficulties, and become self-renewing virtually forever."[470]

Mankind had faced problems and challenges before. If only the human mind could be allowed to soar freely – through freedom from discrimination, waste and poverty – there was a good chance that the present battle could be won. Better technology could substitute immaterial design for matter and thus lead to a much more efficient use of the given resources. Thus even unlimited economic development was not impossible, since constantly improving technology could make more out of less. "Thorough innovation of our economy", he ended *Riches to Rags*, "is the only route toward the future which is not impossible or absurd. The earth may have its limits, but the domain of the human mind is virtually boundless – if only it is free from self-imposed bondage to the habits of thoughts of past ages."[471]

In this way it is apparent that the seeming paradox in Dovring's thinking is to a great extent resolved. The wastefulness built into the social structure could actually be seen as a barrier to true progress, and what seemed like an old-fashioned way of thinking was needed to censure the appropriate shortcomings and remedy them. Thus traditional morality was an excellent companion to the modern soci-

468 Dovring (1988), p. 1.

469 Dovring (1984), p. 137.

470 Dovring (1984), p. 171.

471 Dovring (1984), p. 171.

ety as it prepared to break its final shackles and evolve into the full Optional Society.

Clashes with mainstream economics

The path of any reformer must sooner or later take him into the realm of economics. Dovring's radical ideas attacked the fundamental workings of the economy and thus made theoretical claims to describe how the reality of this economy and its functioning were rigged. The changes he advocated were not just taken out of thin air, but based on a coherent way of thinking, which often took the shape of economic theory. He was firmly rooted in institutional theory and much of his thinking can be placed in an American tradition going back to the young scholars who crossed the Atlantic to study economics in Germany in the 19th century. This way of thinking had been very influential in the past, finding its niche in reaction to the most obvious flaws of neoclassical theory, but began to lose ground as the Keynesian school emerged in the 1930s.[472]

As both a professor of land economics and as a radical thinker on societal reform, Dovring was bound to come into conflict with mainstream economic thinking. Indeed, the way economists viewed the world was part of the problem. Economics itself had to be reformed. Keynesianism had gone out of fashion in the 1970s, but the static, atomistic principles of neoclassic economics seemed to encourage both waste and laissez faire policies thereby helping to maintain the destructive tendencies of American society. Therefore economic thinking had to be reformed in a number of ways and the institutional school again foregrounded as a guiding star for societal reform. This aspect of Dovring's life thus highlights an important ingredient in the modern debate – is the dominant way of thinking about economics good enough, or will it have to be reshaped or even discarded in the close future?

This section will deal with Dovring's clashes with mainstream economics. It begins with a brief discussion of neoclassic and institutional theory.

A brief history of neoclassical and institutional theory

Both neoclassical and institutional economics emerged from the thoughts of those known as the classical economists. Men like Adam Smith, David Ricardo and John Stuart Mill produced writings that share several aspects. They were primarily interested in the question of long-term growth and the share of the total production of the economy that befell not different individuals but groups in society. Reactions against the previous era of guild and government regulations led them to adopt a highly liberal view of trade and the workings of the economy. They were also usually pessimistic, believing that growth had a natural end due to

472 Carlson (1995), p. 94.

the law of diminishing return. This led the historian Thomas Carlyle to coin the term "the dismal science" to denote the budding field of economics[473], a description that has stuck even though later developments would rather transmute economics into a science of unrestrained optimism.

The emergence of neoclassicism had been a long time in the making. The reasons for its sudden burst of success in the 1870s have not been fully investigated.[474] It is reasonable, though, to assume that theoretical difficulties within classical thinking, fear of socialism armed with Marxist economics – in many ways a branch of the classical school – and enthusiasm for mathematics as a tool had much to do with it. A number of special features distinguished it from classical economics. The long-term perspective used by the classical economists was abandoned in favor of analysis of the allocation of resources here and now in a given market. The influential economist Joan Robinson famously joked that this meant that interest shifted from the big picture of the fate of the future society to questions like "why does an egg cost more than a cup of tea?"[475]

Further, the main unit of analysis ceased to be a societal group – such as laborers, capitalists and landowners – and became the individual. The workings of the economy were then derived from the behavior of the individuals who functioned through calculations of utility. Norms for action, habits and legal frameworks that might shape the choices of individuals did not enter the picture. The belief that the law of diminishing return would mean the end of growth disappeared, as the optimism of the late 19th century was incorporated into the system.[476] Indeed, neoclassical economics did not even bother about growth and development over time, but saw the market as allocator of resources in a single time period to be the overwhelmingly most important issue. A cursory glance therefore makes it apparent that Dovring would come into conflict with any theory based on such principles.

The roots of institutional thinking lie in 19th century German thinking. Classical economics of British origin with its insistence on free trade and liberalism did not fit well with German traditions. It was also felt that British economics would not benefit the specifically German situation, since the latter was in the process of industrialization and so had to follow other principles than fully industrialized Great Britain. German economic thinking focused on society as an organic whole and assigned important guiding functions to the state. It was also generally skeptical of free trade – at least before industrialization had been completed – and centered more on the nation as an important unit of economic progress.

It is obvious why the German school appealed to many American economists

473 Barber (1970), p. 68.

474 Sandelin – Trautwein – Wundrak (2001), p. 96.

475 Barber (1970), p. 165.

476 Barber (1970), p. 164.

as an alternative to British classic economics. Here was a young country with aspirations of becoming industrialized and in this process the *laissez-faire* and free trade of the British school seemed to many to do more harm than good. Between 1820 and 1920 close to 9 000 Americans crossed the Atlantic to study in German universities.[477] It was felt that the conditions of Germany more resembled America and that German knowledge was more applicable to the American situation.

It has been said that two typically German traits in particular had a great influence in the way American thinking developed: a tendency to view economics in a highly normative way and a tendency to focus on the practical usefulness of theory.[478] The result was the early introduction of eagerness for reform in American economics. This was apparent in many of the so-called pre-institutionalists. Richard Ely for instance, the "human dynamo" who had a huge impact on American economics, preached a sort of Christian Socialism and left the secretaryship of the American Economic Association in protest after this veered away from "quasi-religious social reform" – only to return in triumph seven years later as he was elected as its chairman.[479] The event suggests just how mainstream radical reformism had become in American economics.

Such traditions produced the current of thought called institutionalism, whose most influential proponents were Thorstein Veblen and John Commons. The individuals grouped as institutional economists were a diverse collection of thinkers who focused on entirely different things. There shared, however, some common features, among which the skeptical attitude toward the individualistic theory of the neoclassics is a cornerstone, together with the view that the existence of institutions has a huge influence on the decisions of individuals and firms and thus on the workings of the economy.[480] Eagerness for reform and radicalism were also prominent in many institutionalists. Veblen, for instance, hoped for a sort of socialistic society operated by technicians and engineers.[481] In the company of the American economists of late 19th century and early 20th century, then, Dovring would not have stood out as especially radical at all. In this context, neoclassical theory must also be viewed as the economics of adherents of the status quo.

Institutional economics climaxed in influence in the 1930s and then declined. This was most likely due to the success of Keynesian thought, which became both an alternative to neoclassical thinking and the main source of guiding theory for government policy.[482] Perhaps the message of a possible harmony in capitalism spread by Keynes also made his theories more palatable than the widespread radicalism usually associated with institutionalist thinkers.

477 Carlson (1995), p. 45.
478 Carlson (1995), p. 48.
479 Carlson (1995), p. 52.
480 Carlson (1995), p. 91.
481 Carlson (1995), p. 74.
482 Carlson (1995), p. 94.

The total result of the influence of the institutionalists is a matter of some discussion. Many think that institutionalism never succeeded in founding a coherent theory of economics in itself, but nevertheless made lasting contributions to economics by stressing the importance of areas that needed special focus. Thus many fields of applied economics were greatly developed by institutionalists. The establishment of land economics, for instance, as a separate subject – in which Dovring held a chair – was to a great extent a result of research by institutional thinkers. Indeed, the first text book in the subject was written by none other than the human dynamo himself, Richard Ely.[483] Land economics was then from the beginning a typically institutional field, but it was increasingly infiltrated by neoclassical methods. Dovring's own textbook in the subject – to be discussed in more detail below – published in the mid-80s, laid emphasis on institutional theory and was openly hostile to key neoclassical assumptions. Gene Wunderlich, who reviewed Dovring's textbook, noted that it was probably the last of its kind in the institutionalist tradition of land economics going back to Ely.[484] Almost nostalgically, he wrote that much "of early land economics is now disappearing among its offspring specialties in more sophisticated dress as natural resource economics, environmental economics, benefit-cost, bid-rent functions and contingent values."[485]

"More sophisticated dress" probably should be read as a dress of neoclassical formalism with mathematical laces. If one opens a more modern textbook in the subject, it is obvious that the theory of land economics has been almost entirely submerged in neoclassical thinking and most of its features are expressed in the language of neoclassical models and graphs. The dispersion of institutionalism into several subfields, which actually means the dismantling of its core in favor of more coherent schools, noted by Wunderlich is interesting. It fits nicely with the suggestion made by Richard Ely's biographer Lafayette Harter. He claims that it was actually the success of the institutionalist school in many subfields of economics that led to its fall. It became too dispersed and bereft of a solid centre. Other schools adapted by integrating aspects of its valuable insights and some elements became the study of bordering fields like sociology. Instead of establishing a new orthodoxy, it produced many new insights that the old orthodoxy could incorporate in order to strengthen its claim to be able to describe the future society.[486] This deprived its core of any greater power of persuasion and as Keynesianism emerged it was easily left in the shade. The loss in popularity of Keynesianism has meant something of a renaissance for institutionalism.[487] But it is far from being orthodox and after the fall of Keynesianism, neoclassic theory reigned su-

483 Dovring (1987), p. 13.
484 Wunderlich (1988), p. 491.
485 Wunderlich (1988), p. 491.
486 Carlson (1995), p. 94.
487 Carlson (1995), p. 95.

preme, doing what it did best – namely focusing on the effective short-run allocation of resources – and leaving the big decisions to market forces. Major criticism is now being expressed of the neoclassical orthodoxy and time will tell what will happen in the future with the subject of economics.

In all this we find Dovring, trying to pave the way for his textbook in the mid-1980s. As Wunderlich pointed out he had written it in a tradition that was being dispersed and integrated in neoclassic theory. The controversy that followed is therefore extremely interesting for two reasons. First, it is a piece of doctrinal history – being after all "the last of the institutional textbooks on land economics" – reflecting a battle of methodology which was largely over when Dovring entered. Second, its attack on neoclassic orthodoxy gives resonance to much of the criticism expressed about it today from many different directions – including from within the economics community itself. Therefore Dovring's textbook can be seen both as the end of something and the beginning of something.

The Textbook Controversy

As a professor in land economics, Dovring gave lectures in applied economic theory. His students were supplied with the course material in a mimeographed syllabus written by Dovring himself. When he retired, he decided to find a publisher for it. In 1984 the syllabus ended up at the office of Jay Bartlett at the Breton Publishers office. Bartlett seems to have been a shrewd businessman, but completely uninterested in ideological and methodological concerns. He passed the manuscript to two anonymous reviewers for comments, one of whom – only to be known to Dovring as the hated reviewer no. 2 – was clearly a devoted follower of neoclassic theory, while "no. 1" seems to have been more favorable to institutional theory.

These two reviewers supplied Bartlett with their comments. Upon receiving them he wrote a letter to Dovring in May 1984 in which he used a careful and friendly tone. He declared that he was not an economist, but a mere businessman whose views on the manuscript were derived solely from his "reading of the marketplace". In the pure technical matters he relied fully on Dovring, but asked him to consider all the criticism raised by the reviewers.

The criticism was what could be expected given the differences between institutional and neoclassical theory. Both reviewers wanted a clearer grounding in basic economic principles, but no. 2 lost no time in unfurling his flag and demanded more applied micro theory – something of a hallmark for neoclassics. Bartlett knew enough about the marketplace to sense that it was in this case divided in different orientations, but was probably also aware that no. 2 represented the prevailing view. "I know", he wrote to Dovring, "that many of his positions differ from your own, but he does represent the views of many potential adopters. Note that he seems strongly inclined to forgive the differences, if you'll just give him some microeconomic applications to play with."

This ought to have annoyed Dovring. He had just finished his summation of the wastefulness of society – the *Riches to Rags* book – and was here confronted with a demand to write his own textbook in a way that supported much of the economic thinking he had castigated. No. 2 had actually requested that micro theory should be the "basis of treatment" of the entire book, something that must have appalled Dovring. "In my thinking", he answered a week later, "micro theory used in isolation is a polite term for consumer fraud. To keep micro theory honest it must be subordinated to conceptualization, macro theory and institutional theory, plus a large helping of facts-of-life, all of which is how my text has been thought out and how it is taught to my students."

It is important to note that Dovring did not skip micro theory altogether. The final version of the textbook contains a good deal of it, but the essence of the dispute is that micro theory is not the "basis of treatment" that no. 2 wanted. To use micro theory as a basis of treatment is natural to neoclassics, since it is a theory built on the assumption that the workings of the economy can be derived from its individual atoms. In such thinking micro theory *must* be the basis of treatment. In the final version of the textbook Dovring openly declared his hostility to this way of thinking: "Some textbooks of economic theory maintain that the economy as a whole can be understood as the sum of its many small parts. That is, one studies the micro units and how they make their decisions and this can be summed into a picture of the macro economy. Stated in this simple form, the proposition is a fallacy ... In economics, this only seems to hold when a static analysis is combined with a limited perspective. The micro units are seldom able to act with complete freedom of choice. To a large extent they are constrained by the larger context within which they have to operate and which imposes limits for choice."[488]

Here then – in a nut shell – lay the main difference between neoclassics and institutionalists. Land economics had in the past been a typical domain of institutionalism, but it was now being claimed by neoclassicism. Dovring thought that neoclassical thinking had nothing to offer to the subject of land economics. In his letter to Bartlett he attacked no. 2: "With No. 2's emphasis on micro theory, he should really have dropped the whole concept of Land Economics, because in such an analytical setting the distinctiveness of land – what makes it different from commodities – is reduced to very little."

As to no. 2 and his "ideological friends" forming an important part of the market, Dovring was skeptical. "That group", he claimed, "can not be a strong potential market for land economics. Our core audience is the institutionally oriented land economists." Apparently, then, Dovring had not understood the full extent of the neoclassical onslaught on the land economics that used to be an institutionalist bastion. He did understand, however, that neoclassical thinking

488 Dovring (1987), p. 9.

went well with the resource-wasting policies of *laissez-faire* favored by the Reagan administration. "The scholarly profile of no. 2", he wrote to Bartlett, "is all too apparent. It places him in a league with all the wishful thinking that has given us 'Reaganomics'. When this misdirected experiment collapses (as it is sure to do, well within the next four years), then a great deal of fresh thinking will be called for." He then advised Bartlett to read *Riches to Rags* to learn more about this. Indeed, in the final version of the textbook Dovring declared in the most careful wording he could find – an impressive achievement given his polemic nature – that the methods of neoclassical theory was not good enough to deal with the present state of the economy: "Static or near-static micro analysis projects the appearance of doing a good job in placid economic times – in times when there are few upsetting changes. It does much worse in times when important changes are underway, the times when we need economics the most."[489]

Already in his critique of neoclassical theory, the next point of dispute started to arise. In the letter from Bartlett, the publisher carefully pointed out the fact that the "syllabus sometimes blurs the distinction between normative and positive statements". As mentioned above, this was in good institutionalist tradition in which the highly normative character and practical orientation of the German school had developed into a great enthusiasm for reform. In contrast to this, neoclassical economics has, ever since its inception, tried to portray itself as a highly positive science, aspiring to achieve the same objective scientific status as physics – this is an important reason why it came to rely so much on mathematics. The paradox here is that neoclassical theory in trying to sever all links to political views actually becomes extremely political – it favors the view that politics should be left out of the economy altogether and therefore comes close to becoming the perfect tool for *laissez-faire* politicians. To Dovring on the other hand, theory had no value in itself but only as a guide for practical actions and sound government intervention. "A good theoretical framework", he wrote in the final version, "is an instrument of orientation that is indispensible before our research can become action oriented. For well-directed action, there is nothing more practical than solid theory ... But the framework needs to be comprehensive and this places macro and institutional analysis ahead of micro theory."[490]

Probably the whole tension between positive and normative meant nothing to Dovring – to his nature they were one and the same. In the textbook he wrote: "We will try as far as possible to take society as it is, to explain rather than critique it. In extreme cases, the consequences of one institutional alternative will be so evidently superior to those of another that a practical conclusion can not be avoided. If the current institutions tend toward disaster, this obviously should transcend the usual distinction between the positive and the normative."[491]

489 Dovring (1987), p. 10.

490 Dovring (1987), p. 10.

491 Dovring (1987), p. 12.

This meant that he felt fully entitled to include many of his conclusions on how the American society should be reformed. He devoted the last chapter of the textbook to the subject of government intervention and declared as a matter of fact that the need for government control over resources had increased. This was brought on partly by the increasing complexity of the economy – a fact that is most often used to discourage government intervention – but mostly by the "greater scarcity of some resources because of the unprecedented quantity development of modern economic systems."[492] He also attacked the short-sighted optimism of the general public and as usual brought up the problem of oil dependence. The solution to many of these problems through the farming of biomass that he advocated in *Farming for Fuel*, was integrated into the textbook. His optimistic side was also revealed and he prophesied that the use of artificial photosynthesis to produce energy was possible in the future. "The economics of this", he wrote, "is not likely to be known until the next century."[493]

Dovring's textbook, then, was indeed filled with his own views. Bartlett was surprisingly forgiving about this and made it clear that he did not want Dovring to mask his own opinions. He did, however, demand that these be put into perspective by clearly labeling the opinions as opinions and not matters of facts, or by "including other alternative opinions alongside your own".

This was not acceptable to Dovring. Nor was he pleased by the degree of "monitoring" suggested by Bartlett. "I am puzzled", he wrote, "by the several levels of supervision foreseen in you discussion: You personally would not interfere with decisions on professional matters, but your company head apparently would have to a say as to how far I have complied with reviewers' recommendations, and you anticipate a second review by these same reviewers. And then there will be an editor to boot. I am not sure that I am prepared to submit to so much monitoring, after the many books and the load of other writing (in this country and across the world) that I have below my belt."

He did however produce an expanded manuscript which was sent to Breton Publishers. Ed Francis, the head of the company referred to by Bartlett, had intervened to deal with Dovring. In October 1985 another pair of reviews were sent to Dovring. Dovring was very annoyed to find that once again no. 2 had been the reviewer that the company put their trust in. Francis declared that "more basic principles (microeconomic concepts)" was needed. To Dovring that was just dull repetition of the last dispute and once more the schism between neoclassics and institutionalists was brought to light. "I would not write my book on the specifications of this second reviewer", he answered in November 1985. "My professional reasons for this are strong and are spelled out in Chapter 1 of my manuscript ... There are many suggestions in this second review which I reject on professional grounds." Once more he also proved to be unable to grasp the neoclas-

492 Dovring (1987), p. 500.

493 Dovring (1987), p. 191.

sical wind blowing in the lecture halls: "You remark in your letter that sales will depend on giving faculty what they believe is best for their class, that argument cuts both ways. Many of us, myself included, would not adopt a book written to your reviewer's specifications ... Academic opinions are divided, and I know where my following is. To bad it could not be represented among your reviewers this time."

By now the publisher's interest had faded. They now tried to get rid of both Dovring and his textbook. Somewhere along the line, however, a contract had been signed. The publishing house was obliged to publish the book if Dovring wanted it and he did. They tried to get him to take his manuscript to another publisher, but he refused and referred to the contract.

The book was published. No. 2's suggestions were disregarded. Dovring had won a victory, but it was a hollow victory. The print run was limited and the minimum amount of marketing was devoted to it. It made no mark whatsoever on either the subject of land economics or the methodological debate. In 1989 54 copies had sold. Instead, the leading textbook in the subject became a revised version of one written earlier by Raleigh Barlowe. Its 1986 cover displayed not a picture of something connected to land in real life but a neoclassical graph depicting average and marginal cost curves for optimizing land output. Here was indeed a book that no. 2 would have approved of, that incorporated neoclassics and made great use of the microeconomic concepts that were lacking in Dovring's textbook. Barlowe's book took the place that Dovring had planned for his own book. Wunderlich's nostalgic review of Dovring's textbook suggests that by this time the subject of land economics had indeed been become immersed in neoclassical thinking.

This did not mean that Dovring was beaten. The textbook was just part of the much grander plan of reform already embarked upon when he entered the textbook conflict. Neoclassical theory as well as obsolete assumptions that coursed through the entire structure of economic thought supported the dangerous policies that threatened society itself. "By praising the market in equilibrium", he wrote, "neoclassical analysis simply arrives at paying homage to the economic policy which happen to be in power at the time."[494]

In a number of books all subtitled *The Political Economy of...* he tried to attack the theoretical framework of mainstream economics and to draw from his theoretical thinking the correct conclusions needed for a sustainable society.

What follows is a brief account of this struggle.

Serious economics

When it came to economic theory, Dovring was adamant that it had to be attuned to the present risks in the modern economy. "Serious economics", he wrote,

494 Dovring (1987b), p. 154.

"is about the future."[495] Mainstream economists locked themselves in their ivory towers, thinking about the economy with the conceptual tools created by people in a bygone age with a totally different situation. Their theories were much too short-sighted and as a result could hardly cope with the task of guiding society into a future with greater scarcity of resources.

A central problem was the way productivity was measured. In *Riches to Rags* he noted the fact that increase in waste could be counted as increase in productivity through increasing the G.D.P. index. In an effort to come to terms with this problem he made an attempt to contribute to the theory of production.

Theoretical problems when it came to measuring total increase in economy productivity as distinct from increase in productivity in a segment of the economy had been raised by some economists. There was also the question of how the productivity of different factors could be aggregated into a single measure of total productivity. Dovring was concerned that the gains from increases in productivity – one of his favorite topics – were not spread evenly over the population. Most important, perhaps, were the factors taken into account as production factors in economic theory. Dovring, who always had the word scarcity in mind when dealing with economics, was aware that a resource had to be scarce for anyone to assign it an economic function. No one would dream of assigning an economic function to air, since it was an abundant resource and could therefore be taken for granted. Only scarcity meant that a resource had to be the object of economic theorizing. When the classical economists, then, laid the foundation for modern economics they identified three factors of production: land, labor and capital. This reflected the situation of scarcity the economy faced at that time. By now, however, the situation had changed and the list of production factors taken into account in economic theory had to be supplemented with energy and human capital.

In his book *Productivity and Value* published 1987, Dovring tried to both supplement the traditional list of factors and deal with the problems of measuring productivity. Energy had to be integrated into economic thinking as a concept fully comparable to key concepts such as production, consumption and investment. Indeed, energy should be "regarded as a separately recognized and analyzed factor of production with its own productivity ratings and productivity variations."[496] This would mean a very different conception of the changes in economic activity that has taken place in the last 100 years. For instance, productivity measurements in agriculture when measured in productivity per unit of labor show an extreme increase in productivity. On the other hand if measured in energy productivity, the changes in agriculture shows a loss in productivity due to the enormous amounts of energy used to produce fertilizers.[497] The end of what

495 Dovring (1987b), p. 76.

496 Dovring (1987b), p. 73.

497 See for instance Jansén (2005), p. 218.

Dovring called the Petroleum Age with its abundance of cheap energy would mean that greater interest would have to be paid to energy in economic theory and analysis.

The same went for human capital in the shape of management and design. The failure of the classical economists to take these into account could be understood because of the low complexity of the society around them. The important role of management "was largely overlooked because in those days of industrial simplicity, managerial talent was in chronic oversupply, thus it could be treated as a free good analogous to sunshine or to the health of workers when healthy workers were in excess supply."[498] With increased complexity, management had become a scarce resource and its omission from economic theory had become a problem. Wasteful use, through racism and discrimination, could be more easily overlooked when mainstream economics viewed it as an abundant resource.

The view of capital also had to be deepened to take the human mind into account. Human ingenuity was the motor that changed productivity and it manifested itself in the way capital was reshaped for more effective use. The essence of capital then, was design. This made it "distinct from all three primary factors of production. Designs are for the most part human-made; we can say that here is the source of productivity change. Land, labor and energy as natural facts cannot be attributed any ability to improve in productivity."[499] Capital in this sense was the "sole active source of productivity improvements" and such a complex phenomena that it could not be treated as a single concept. This complexity of capital had to be acknowledged in economic theory.

When it came to the problems of measuring total productivity increase as distinct from increase in a single sector of the economy, Dovring proposed a mathematical solution. This solution actually did mean that the increase in productivity in a single sector of the economy was preserved as the level of aggregation increased. He had proposed the same solution in a paper in 1979 and earlier another economist had proposed the same solution but in more cumbersome mathematical form in 1961.[500] For what it was worth, then, Dovring actually did make a contribution to the field however small.

These observations pointed to the flaws of total factor productivity which contained too many assumptions that he was unwilling to make. For instance, the prices or unit costs of the inputs were held constant for more than one period. In reality not only the prices themselves but also the relative prices among the input factors themselves changed. To assume that they did not was to miss too much of what was going on. Besides prices were not an ultimate guideline to the relative scarcities of the factors, since this had more to do with the power structure in the economy than with the actual factor endowment. A disaggregated approach was

498 Dovring (1987b), p. 122.

499 Dovring (1987b), p. 106.

500 Dovring (1987b), p. 106.

a possible way out, "a fallback position" from the pitfalls of total factor productivity. This meant that the productivity of a single productivity factor was to be measured and not in terms of nominal value but in the quantity of a single homogenous output.[501] Capital was far too complex to be treated in this way, but he claimed that human labor was not. "Exploring how much labor (of all kinds, without any quality weighting) that goes into the production of a unit of goods, will tell us something essential about economic progress in the system."[502] This accumulated single-factor productivity, he claimed, would bring a new sense of clarity to the whole economic system. Total factor productivity could obscure the actual distribution of the gains from productivity increases, but notaccumulated single-factor productivity together with consumer surveys "for selected levels of family income, and this will then show if there is any difference in consumer welfare, deriving from productivity improvement, as between the income strata in society. Is it true that some classes were more favored by others? The answer will be there to find out."[503]

But the effects of the way productivity was measured had even further effects on the economy of the real world. When measuring productivity in nominal terms, the true contributions of the factors in physical terms could be disguised in favor of economic inequality. If the use of capital was lessened and labor increased, this meant that the measure of production for capital increased and decreased for labor. This was because of the law of diminishing return that stated that the marginal increase in output decreased with increased input of a specific factor. Given such a change then, low wages for labor would be quite justified because of the assumption that wages were contingent on the value of its marginal product. "The circle closes", Dovring concluded, "the analysis serves to justify the status quo rather than explore how it might be changed."[504] He referred to this procedure as "a game of class laws", that could be demolished if the productivity of the factors were measured not in value but in physical output. "Labor cannot by weighted in advance by its wage, nor capital by its book value."[505]

This interest in inequality was only the beginning of another project. In *The Optional Society* published 1971, he had praised the tendency in capitalism to reduce inequality. Indeed, there were plenty of statistical data to support such claims. Since then, however, the trend had turned. The total wealth of society continued to increase, but with increasing economic inequality. This trend is continuing to this day, not only in the United States.

In 1991 Dovring, now 75 years old, published *Inequality – The Political Economy of Income Distribution*. Writing it was the natural development of both the

501 Dovring (1987b), p. 159.
502 Dovring (1987b), p. 160.
503 Dovring (1987b), p. 191.
504 Dovring (1987b), p. 154.
505 Dovring (1987b), p. 154.

course his reform project had taken and the course of American society. The tendency toward greater equality that had been a key feature in the road to the Optional Society had disappeared. Dovring was by no means an advocate of total equality in the communist sense. On the contrary he was well aware that a functional inequality was necessary, but the problem was that inequality could get out of hand: "There must be some inequality of income because there has to be some motivation to excel and to seek the kind of occupation where one does one's best. However, too much inequality will hurt society as well as many individuals if economic advantages are too concentrated, leaving many people without any realistic hope."[506]

At its core, then, the problem was the same type of waste of human talent that he had lamented earlier. The circumstances of unprecedented wealth in the world made the problem even more alarming: "In remote times, penury of the masses was simply the general human condition, from which the luxury of the ruling class was an exception possible only for a small minority. In the industrial age, the problem is different, since enough is produced that no one really needs to live in penury."[507] This meant that the increased inequality in America could hardly be natural, but was the result of the built-in wastefulness in the economy and the free-market ideology adopted in the 1980s. He had said it before and said it again: the best economy was a mix between market and government guidance. The totally free market had never existed in the way its adherents imagined it: "The only perfect market that ever existed was and is the jungle, the wilderness void of human interference. All else is or has been modified by culture ... Never has competition been completely unhampered, and neither could there ever have been any economic relations between people without some institutional rules telling them how these relations are to be handled."[508] The conclusion was obvious: "The ideal of complete freedom from societal interference is, therefore, an unreal construct – a myth. Without any institutional framework, an economic system could not function at all ... Any meaningful debate about societal control versus economic freedom must, therefore, be about a mix of public arrangements versus individual freedoms – it cannot be about an either-or situation. It is from such a perspective that one can discuss the economic policies of the 1980s."[509]

Laissez-faire then was nothing natural, but simply just another form of government policy. It was also a bad government policy. The Reagan administration created a large government budget deficit by lowering taxes and the overall economic results were dubious: "In net terms it is uncertain whether the country as a whole has made any progress at all during the 1980s. Rising real income for the rich and the well-to-do has been obtained at the expense of falling real income for

506 Dovring (1987c), p. 2.

507 Dovring (1987c), p. 106.

508 Dovring (1987c), p. 127.

509 Dovring (1987c), p. 127.

the lower-income members of the population, who often have to live with the most of the environmental degradation as well."[510]

He attacked many of the rationales that increased inequality would mean higher economic efficiency. The assumption that savings would increase with higher inequality, because it was harder to consume a higher income in full, was refuted on the grounds of conspicuous consumption. This was a term that Veblen had introduced, that meant that instead of saving more when income increased, people engaged in buying expensive luxuries to enhance social status. The tendency in America to adore the rich and want to emulate them made this a more likely outcome from increased inequality than a rise in savings.

A very interesting twist on the question of taxes was that lowering taxes to provide incentives for increased economic activity could actually be wasteful. Dovring turned Reagan's statement that in his acting days he would have done five instead of four movies a year if the tax rate had been lower on its head. The same argument as Reagan put forth is usually made in textbooks in economic theory and therefore Dovring's twist is extra interesting. Here he proved he was still, aged 75, capable of brilliant remarks: "the proposition is in fact an excellent argument in favor of the graduated income tax, with high marginal tax rate on high incomes. The country and the world did not need five or more Reagan movies in a year, four were quite enough. Instead, after satiating its appetite for Reagan movies, the public would need movies of some other style and content: more variation in the sphere of cinematic culture. This would also have meant more of a chance for other talent, besides what was already sufficiently recognized, to make its contribution to the cultural marketplace."[511] In this way higher taxes could actually increase the abundance of the market, thus also increasing the numbers of options – to the end he kept the Optional Society as the model society to strive for.

In sum then, Dovring's books on economic theory do contain many brilliant insights and suggestions. On the other hand, they are not crystal-clear in argument and style and much of the impact is lost through confusion and weak argumentation. Reading them is like searching for pearls in murky water – there is often haziness and splashing around, but suddenly there it is – a pearl of shining insight, well worth the trouble. It is not surprising, then, that most of the books received poor reviews. Still, taken as a whole – together with his thoughts on the Optional Society and waste – there is a solid structure of theory here beneath the hazy prose. It is this full structure of Dovring's mind that should be reviewed, not pieces of it. Given the development of the world, it is impossible to dismiss him on account on details about his prose and disposition of arguments. Serious economics, indeed, really does seem to be about the future.

510 Dovring (1987c), p. 125.

511 Dovring (1987c), 128–129.

Dovring and the environmentalist movement

But if Dovring was so concerned about the future and the use of scarce resources, why then did he not join forces with the wave of environmentalism that engulfed America during his own lifetime? That question is even more relevant, when we consider the isolation he and his ideas faced. Could he not have reached out with better impact if he had channeled his thoughts through the infrastructure of an existing social movement?

First of all it must be said that the history of environmental ideas is far from clear. A number of studies have appeared with different conclusions on the nature of these ideas, their origin and development.[512] Some have argued that environmental awareness is a remnant from past pagan times, others that it is the result of new scientific ideas in biology or simply a reaction against the careless use of nature in modern times. Environmental thoughts indicate to some a deeply conservative approach in the sense that nature must be conserved from modern society. Others instead see something fundamentally radical in the way early environmental thinkers in the 19th century approached nature.

It is tempting to see an early rift here. What is common ground for all environmental thinkers is – obviously – their concern for nature in its encounter with modern society. But what is nature and how is mankind to approach it – as a resource to be used wisely in order to achieve long-term utility, or as a thing in itself with inherent values that must not be tampered with? In this question may well lie a great dividing line in all environmental thought. The early American effort to preserve national forests gives a good example. It was the result of a simple conclusion: forests would be eradicated if nothing was done to control their exploitation. In this John Muir and Gifford Pinchot stand out as representatives of the divergent streams in such a simple insight. Muir was a Scottish immigrant from the countryside, the very personification of American wildlife, co-founder of the still active environmental society The Sierra Club.[513] Pinchot was a forester of patrician stock who became highly influential in American national policy. Both were worried over the destruction of the forests, but for different reasons. Muir maintained that nature had an inherent value in itself and was the source of endless joy for the human mind. "We are now in the mountains and they are in us, kindling enthusiasm, making every nerve quiver, filling every pore and cell of us", he wrote in his diary during a stay in the Sierran mountains. "Our flesh-and-bone tabernacle seems transparent as glass to the beauty around us."[514] There was a fundamentally religious feeling in being out in nature. Now such experiences were threatened by logging companies and only government intervention could

512 See for instance Bramwell (1989), p. 22–23.

513 See the homepage of The Sierra Club: http://www.sierraclub.org/john_muir_exhibit/, 19.11.2007.

514 Muir (1997), p. 161.

stop it. For thousands of years "God has cared for these trees ... but he cannot save them from fools – only Uncle Sam can do that."[515]

Pinchot felt the same pressing need for government intervention. But he had other reasons for it. In fact his writings are in many ways similar to Dovring's. "When the natural resources of any nation become exhausted", he wrote in 1910, "disaster and decay in every department of national life follow as a matter of course."[516] He worried about the national supply of coal that was going to run out within 50 years at current consumption rates. The wastefulness in extracting the coal made things worse and the "waste in use is not less appalling".[517] Here was the same distaste for waste in itself that Dovring expressed, together with the overall concern for national welfare and future prospects. This went for the forests as well, which should be managed through sustainable levels of exploitation. In the end it came down to utility: "Conservation", Pinchot wrote, "means the greatest good for the greatest number for the longest time."[518]

These two views could only coexist in the same movement for a given time. Should forest reserves be withdrawn from use altogether or commercially exploited in a sustainable matter? Muir advocated the first, Pinchot the second and a rift in the conservation movement was a fact. From Muir stemmed what is called preservationism, while Pinchot's line moved forward under the name of conservationism. Muir remained president of the Sierra Club until his death in 1914 and his impact should by no means be underrated. But it was the conservationism of Pinchot that became dominant. With Theodor Roosevelt as president from 1901 it became official national policy. In 1905 Pinchot became the first executive of the newly started US Forest Service and managed to ensure its control over the forest reserves.[519] "Conservationism has captured the Nation", he wrote in 1910.[520] During the decades that followed, however, it started to run out of steam. The cravings of the forestry and mining companies became harder to resist. For instance, the suggested ceiling on timber extraction at the Willamette Park of 754,000 cubic meters, was exceeded in 1952 and was twice as large in 1962.[521] In 1954 Grant McConnell at the University of California, wrote that conservationism had been "the most conspicuous cause on the American political scene. Today, however, the movement is small, divided and frequently uncertain."[522]

515 Muir (1997), p. 720.
516 Pinchot (1967), p. 4.
517 Pinchot (1967), p. 6–7.
518 Pinchot (1967), p. 48.
519 Nash (1967), p. xvi.
520 Pinchot (1967), p. 132.
521 Hughes (2001), p. 280.
522 McConnell (1954), p. 463.

When Dovring settled in America, the conservation movement, then, was largely irrelevant. Large scale concern over nature was instead to be propagated by the modern environmental movement. This was ignited by the 1962 publication of Rachel Carson's *Silent Spring.*[523] Its title referred to the silence resulting from the death of all birds due to the use of pesticides in agriculture. Here, then, was the recognition of the workings of the ecosystem, that chemicals used in one level spread to other levels of the system – birds ate insects and were harmed by the pesticide. Many modern environmentalists view this ecological insight as something that separates the old conservation movement from modern environmentalism.[524]

Carson's work seems to owe little to the conservationist approach taken by Pinchot and his followers. On the other hand she was very much inspired by the panegyric nature writings of Henry David Thoureau and Richard Jeffries, which are much closer to Muir in their view of nature.[525] Indeed, the utilitarian conservation ethics propounded by Pinchot implied that nature was mainly a resource to be used for the good of mankind. Carson stated in *Silent Spring* that the very idea of control over nature belonged to the "Neanderthal age of biology and philosophy, when it was supposed that nature exists for the convenience of man."[526]

The fledgling environmental movement was closer to the preservationism of Muir while at the same time it took on a specifically international outlook. Recognition of the workings of the ecosystem and the view of nature as something with its own value was joined by the awareness that substances and toxins could be dispersed over large distances. It was discovered, for instance, that the DDT that Carson had warned against could be found in the fat of the penguins of Antarctica.[527] Modern environmentalism, as a consequence, was highly international from the beginning – it had to be given the nature of the problems it focused on. It derived much of its concern from the scientific field of ecology, stressing the fact that all parts of nature were interconnected.

Is the answer then, quite simply that Dovring was unable to ride the wave of modern environmentalism because he was an old school conservationist with a fundamentally different stance towards nature? There are indeed striking similarities between Dovring and Pinchot – their views on waste and the risk it posed for the future of the nation to name just one. But that does not fully answer the question.

We need to approach the issue of radicalism. It lies close at hand to suggest that the preservationism of Muir was more radical than the conservationism of Pinchot – after all it wanted to exclude nature from commercial use altogether, a

523 Hughes (2001), p. 15; p. 264.
524 Hughes (2001), p. 265–266.
525 Brooks (1989), p. 7.
526 Carson (1970), p. 261.
527 Hughes (2001), p. 265.

difficult concept to reconcile with a national policy of economic development. The more moderate conservationism prevailed in practice. This suggests a continual problem: how radical, in light of the practical situation, should the movement be? Many in the budding environmental movement found themselves hampered by the moderate approach of older conservation organizations and created their own – only to face the same dilemma when some of their members wanted to go even further. The environmental organization Greenpeace is a good example. Some of its key founders, the engineer and Quaker Jim Bohlen and seaman Paul Watson among them, started out as active members in the Sierra Club, but a breach occurred when the government was carrying out nuclear tests in the Pacific. Bohlen thought it was "helluva good idea to take a ship up to the Aleutian Islands and protest against the bombing".[528] The Sierra Club objected to this, "preferring a more neutral role as simple conservationists."[529] Greenpeace was founded some years later, in 1971, carrying with it the same inherent tendency to diverge. In 1977 Paul Watson left Greenpeace and founded Sea Shepherd to be free to pursue more aggressive methods, but officially "because he felt the original goals of the organization were being compromised, and because he saw a global need to continue direct action conservation activities."[530]

Now let us return to Dovring. How radical was he? In *The Optional Society* he castigated not only the existing societal structure but also the dropouts from it. He held the hippies in distaste because they pursued a lifestyle outside society. This might not seem very radical. On the other hand he attacked the automobile and all that went with it. He declared that a complete break with the past was necessary, that a war economy had to be initiated to avoid disaster. This is very radical. The remnants of the conservation movements would hardly accept such a stance and he probably found them far too moderate to deserve attention. At the same time Dovring was out of tune with modern environmentalism for two main reasons. The first has been touched on earlier: he still held the old-fashioned view on resources that Pinchot had and was probably unable to approach the problem in a more "environmental way". The second is more complex: he did not stress any particular aspects of the environmental problems – he did not dwell on the killing of whales, or the level of DDT in Antarctica. Instead he focused on the system itself, how it worked, how it should be changed. Changing the system is indeed harder than correcting one of its effects – if we see whales being killed we can strive to stop it, if we find that DDT is harming the world we can abolish its use without ever questioning the system that keeps creating these problems. This also became an issue about attracting attention: the Greenpeace leaders became masters of the art of attracting media coverage by concentrating the public mind

528 Brown – May (1989), p. 8.

529 Brown – May (1989), p. 8.

530 See Watson's official biography on Sea Shepherd's homepage: http://www.seashepherd.org/crew-watson.html, 20.11.2007.

on suggestive pictures such as that of bleeding seals. Since the organization became dependent on gifts from the public this tendency was strengthened. Dovring did the opposite: he focused on the system and how to change it, making himself impossible among more moderate conservationists as well as painting a fundamentally different picture from the one offered by modern environmentalists. His radicalism could not be conjoined with that of mainstream modern environmentalism. The result was isolation and lack of impact.

An interesting question raised by this is whether Dovring, with his old-fashioned view on resources and critique of the system, was, viewed as a whole, more or less radical than mainstream modern environmentalism. After all, is there not something highly non-radical in focusing on specific aspects of environmental problems, such as the level of a certain substance or the destruction of a certain coral reef? Is there not a risk that the underlying problem is ignored when great focus is put on its surface phenomena? Dovring instead tried to cut to the heart of the problem – changing the system itself. Perhaps he was too radical for modern environmentalism, especially after this became dependent on widespread popular support. After all, he advocated what could not possibly be as popular as the concrete, often heroic exploits of modern environmentalists – root and branch change of the modern Western way of life. Perhaps Dovring was too radical not just for traditional conservationists, but also for modern environmentalists and thus bound for isolation.

Appendix 5
On light and shadows

By Gabriel Söderberg

Dovring's last book – *Knowledge and Ignorance, Essays on Light and Shadows* – is totally different from everything else he ever wrote. It was published the same year that he died – 1998. He was 82 years old and had reached the stage in life when one feels the need to sum up the years that have gone by. Some would do this by writing a memoir, but to Dovring it was natural instead to share with others the conclusions he had made after a lifetime of scientific work. "More than half a century", he wrote, "has given me ample time to reflect on the limits of our intellectual endeavors. My lifetime of search and thinking has led me traveling across countries and continents and civilizations, as well as across disciplines and problem areas."[531]

He described the path he had followed: "Growing up in the family of a famous poet in the middle of an old-fashioned peasant country (Sweden), I experienced nature more intensely than do most modern people. The vagaries of family history got me started as a historian of the Middle Ages. Breaking this mold, I first veered over into economic history and then, by way of U.N. service, to agricultural economics. This environment let my early experience of nature come to renewed life. The total exposure of intellectual cross-currents has been breathtaking at times."[532]

Now it was time to sum things up, to chalk the boundary lines of human knowledge and to explain his thoughts on how we should deal with what lies beyond us. "During the whole journey of life", he wrote, "the problems of knowledge and ignorance have followed me like a shadow. In formulating, at long last, the conclusions of my labors on these weighty matters, I feel humbled by the power of the subject itself."[533]

Indeed, only now, as he plunged into the full depths of the world, crossing into virtually all fields of human knowledge, did he become truly humble. The attacks against him by geologists and productivity economists claiming that he dabbled in things he did not fully grasp may have been accurate. But now he fully confessed himself to be a layman as he entered the domains of natural sciences such as physics, biology, chemistry and cosmology. He had reached the

531 Dovring (1998), p. vii.

532 Dovring (1998), p. vii.

533 Dovring (1998), p. vii.

point in life when we are no longer afraid of speculation, when we return to the boldness of youth where thought is undisciplined by scientific method and unspoiled by dogma. But if he was ready to accept that there were things he could not fully understand, so should modern scientists as well. He was troubled by their tendency to assume that all that could not be proven did not exist.

Approaching the universe as a seeker of truth, was like being inside a lighted room at night. The lighted room embodied the things that are known to us, those that our scientific methods could reach. "The light provides easy use of the room and its contents."[534] But outside darkness reigned and from the window very little could be seen. There was a way however to see more. "To see what goes on outdoors, dim your indoor lights. Better still, put them out altogether. Adjusting your eyes to the faint light we call 'dark', you will discover many things to be visible out there. The world of out-of-doors is much larger than your room or even your whole building."[535]

In this way, what we did know could actually hinder us when we approached those things lying outside of the realm of scientific methods. Concentrating on things inside the lighted room made the eyes incapable of seeing things outside. He himself had often remarked earlier that modernity was superior to tradition, but his way of viewing economy and culture had been highly influence by traditional thinking all along. As an old man he acknowledged that there was a great deal of value in tradition and that the modernistic world-view failed to see this. "During the Age of Reason in the 1700s, people rejoiced so much in the newfound light of science that they forgot about the kinds of wisdom still dwelling in the dimmer lights of popular culture. Thus was born the intellectual tradition that culminated in the self-blinded arrogance of early materialism: What cannot be proven does not exist."[536]

This blinding of the eyes also caused many other areas of the vast outdoor areas to be lost. The richness of nature could never be fully grasped by scientific methods, nor could much of the spiritual side of mankind. "Naive scientism tends to believe simply that all that is real can also be made known – be subject to objective scrutiny. This human mind, working correctly through scientific methods, could become master of all there is to be master of. Many of us no longer believe this."[537]

In this way materialism barred the windows and people shut themselves off from much of the wonder of the Universe. This resulted in the conception of a world ruled and created by accident, void of meaning and purpose. But this was just another type of faith, a faith that condemned all other types of faith as absurd. "The scientoid mind tends to forget that scientism is itself a faith, based

534 Dovring (1998), p. 7.

535 Dovring (1998), p. 7.

536 Dovring (1998), p. 11.

537 Dovring (1998), p. 31.

ultimately on assumptions that are not scientific."[538] Faith was indeed vital for existence: "Living without faith would be somewhat like living without beauty. Just try it ... and see if it works."[539] Programmatic atheism was thus just another type of faith, only it stood out from the others as the "least rational of all established faiths".[540] Faith itself, together with hope and love, was outside the realm of objective science. "Yet we have no reason to believe that anyone could live entirely without them."[541]

This stated, it was time for Dovring to explain his thoughts on the Universe. Already in his writings on productivity factors, he had laid much emphasis on the design of capital. It was design that could solve the problems of mankind – matter could be substituted by design, more could be made out of less through clever design. Design was the essence of technology and economic development, that which could ease the material restraints on mankind. But design was even more important. Design was the essence of reality itself and therefore materialism was sorely mistaken. "The drawing is not a material thing", Dovring wrote. "All of this, this abstract design, 'rules over' the use of materials. It is in itself not made of any physical material. It is pure thought."[542] Atoms derived their quality not from matter but from design, for they were made of the same elemental quanta that were just arranged differently to achieve different qualities.

This meant he entered the domain of "*logos,* of predefined patterns of existence."[543] It was basically the old Platonian conception of the surrounding world as incarnations of eternal forms. It entered Judeo-Christian thought early on. The Gospel of St. John opens with the famous statement that before the world there was *logos,* that *logos* was with God and that *logos* was God. Dovring interpreted this as meaning: "The totality of all the *logos* in the world, so the gospel writer tells us, is God."[544]

What Dovring believed about the existence of God and God's connection with *logos* is not clear. What is clear though is that *logos* was the true guiding principle in the Universe and that matter and energy were just things to be arranged according to design. All the designs that were capable of existence and what we referred to as laws of nature belonged to a domain "independent of time, space and quantum."[545] All of them together made up what he called the *toolbox of creation.* "Creating anything viable will depend on drawing tools from this

538 Dovring (1998), p. 45.
539 Dovring (1998), p. 45.
540 Dovring (1998), p. 45.
541 Dovring (1998), p. 45.
542 Dovring (1998), p. 48.
543 Dovring (1998), p. 88.
544 Dovring (1998), p. 89.
545 Dovring (1998), p. 85.

treasury."[546] This toolbox also included those human virtues of faith, hope and love. These were "also preexisting principles always waiting in the wings wherever a planet turns out to be at least somewhat hospitable to life."[547]

This brought him into the question of evolution of life. Despite his belief in preexisting forms he did not doubt the fact that evolution had taken place. Evolution was not theory but fact. He did however attack the assumption that the way life had developed was a result of blind chance, of the work of a "blind watchmaker". This conception was entering everyday life through several successful works of popular science, for instance *The Selfish Gene* by Richard Dawkins. This was the result of "a mechanistic, reductionist view of reality".[548] He had attacked neoclassical economists because they expected everything to be no larger than the sum of its parts. Now he waged the same criticism against popular scientists like Dawkins, who with their "masquerade of spurious knowledge" could not grasp the full picture of reality. The toolbox of creation – the only designs that were capable of existence – was overlooked and it was assumed that matter could arrange itself in some mysterious way. Not so according to Dovring. The odds were actually against evolution, since the lowest life forms were actually superbly adjusted to their environment and more complex life forms were much more vulnerable. "If environment adaption", he wrote, "was all that counted for evolution, then a few equilibrated simplicities might have continued to reproduce over the aeons – if they had been created at all."[549] The most apparent puzzle of them all, however, was the human brain. Already in *The Optional Society* he had stated, that the capacity of the human brain had always been hugely excessive and only as the result of a long historical development could its full potentiality be unleashed. From an evolutionary view this was strange. "What was the use of this precious instrument in the jungle? Could primeval man survive better than the monkey? Nothing indicates it."[550]

Even the advent of life itself was highly improbable. In fact it contradicted the laws of thermodynamics that stated that disaggregation "into simpler elements, releasing energy in ever simpler forms, is the normal thing in the physical Cosmos."[551] Yet here it was then, a universe that was constantly erecting ever more complex forms, constantly breaking its own laws in order to impose order and design on matter and energy. Somewhere along the line life had appeared and through evolution it had embarked on a spectacular journey of development. Even mankind was a step in this development, but not the end of it. "Quite to the contrary", he wrote, "we are an instrument for continued creation. Biological

546 Dovring (1998), p. 85.
547 Dovring (1998), p. 100.
548 Dovring (1998), p. 107.
549 Dovring (1998), p. 112.
550 Dovring (1998), p. 113.
551 Dovring (1998), p. 114.

evolution is continued through the sphere of the human psyche into the creative realm of human culture."[552]

But what lay on the other side, when this development had taken its full course? He did not dare to guess, but he pondered the possibility that the Cosmos itself might in the end reach a level of total unity with logos. The technological advances of mankind were an example of how design could substitute for matter, how *logos* became a constantly greater part of the Universe. "In a finalistic setting", he wrote, "the *logos* of the human soul is the ultimate *cause* of the more or less 'material' stages through which creation had to proceed, logically building up the more complex from the less complex."[553] In the end then, Cosmos might well be developing toward some preset goal that he together with influential astronomer Fred Hoyle speculated might be the advent of an intelligent cosmos. "The totality of the *logos* would then have functioned as if it were a cosmic intelligence."[554]

In these ponderings Dovring was probably influenced by a remarkable thinker called Teilhard de Chardin. This Jesuit and paleontologist, who was present when the Peking Man was excavated, attempted to make a fascinating synthesis of Catholic theology and evolutionary biology. Dovring's reflections on the course of the Universe are in many ways reminiscent of Chardin's. In his book on productivity Dovring actually used Teilhard's concept of the *noosphere* to denote "the sum total of knowledge available to human beings at a stated juncture of history."[555] To Chardin Cosmos was moving towards the Omega point, in which this noosphere converged with matter in a sort of personal Universe. Dovring at least saw the noosphere as something constantly expanding, including even that which is forgotten by humans now living – "all that is known or has been known at some time or other" belonged to the noosphere.[556]

In this way it is evident that the thoughts expressed in *Knowledge and Ignorance* were not something that just emerged in his last year. They had been with him for a long time and were not the result of senile wanderings of the mind or a sudden fear of death. That he finally wrote them down, however, most certainly was connected to the fact that he felt death approaching.

Light and shadows was his last project. It was the most ambitious venture of them all. Reforming the human soul itself was the ultimate task for the prophet in the wilderness, the final foray for immortality before moving on to the unknown.

---

552 Dovring (1998), p. 117.

553 Dovring (1998), p. 121.

554 Dovring (1998), p. 121.

555 Dovring (1987b), p. 121.

556 Dovring (1987b), p. 129.

Authors

Janken Myrdal, Professor of Agrarian History at the Swedish University of Agricultural Sciences, Uppsala since 1994. He was awarded his Ph.D. in Stockholm in 1986 for a thesis on medieval agriculture and technology change. He then continued as assistant professor at Stockholm University as well as researcher and curator at Nordiska Museet in Stockholm.

He has written on widely different periods, such as peasant diaries from the 18th and 19th centuries and Iron Age implements. His research centres on agrarian change in medieval and 16th century Sweden, but his extensive publications also include the general and cultural history of the Middle Ages and subjects such as the Black Death, miracles and popular religion, and illuminated manuscripts. In addition, he has worked with modern agriculture and also on forecasting the future on the basis of historical trajectories. Recently he has turned to global history. He is a member of several learned societies including Kungl. Vitterhets Historie och Antikvitets Akademien (The Royal Swedish Academy of Letters, History and Antiquities).

Gabriel Söderberg, Ph.D. candidate in Economic History at Uppsala University. After the award of a Master's degree in Economics, he focused on the field of Economic History and is now writing a thesis on the application of scientific knowledge in the modernisation of Swedish agricultural technology. He has also taught Economics at the University of Gotland.

References

Literature

Abel, Wilhelm (1957), "Book review: Land and Labor in Europe 1900–1950. A comparative survey of recent agrarian history. With a chapter on land reform as propaganda theme by Karin Dovring." *Zeitschrift für Agrargeschichte und Agrarsoziologie* 1957.

Adkins, Douglas (1975), *The Great American Degree Machine*, New York, Carnegie Commission on Higher Education.

Alfonso, Isabel (2007) (ed.), *The Rural History of Medieval European Societies. Trends and Perspectives*, Turnhout, Brepols.

Barlowe, Raleigh (1986), *Land Resource Economics: The Economics of Real Estate*, 4 ed., Englewood Cliffs, Prentice-Hall.

Barber, E. (1992), "Student mobility", in Clark, Burton & Neave, Guy (eds.) *The Encyclopedia of Higher Education*, Vol. 2, Oxford, Pergamon Press.

Barber, William (1970), *A History of Economic Thought*, Middlesex, Penguin Books.

Bengtsson, Frans G. (1938), *Sällskap för en eremit*, Stockholm, Bonnier.

Berelson, Bernard (1960), *Graduate Education in the United States*, New York, McGraw-Hill.

Berg, Maxine (1996), *A Woman in History: Eileen Power, 1889–1940*, Cambridge, Cambridge University Press.

Blecher, George (2002), "När universitet blev ett företag", *Dagens Nyheter*, August 22, 2002.

Bloch, Marc (1953), *The Historian's Craft*, New York, Knopf.

Bloch, Marc & Dauvergne, Robert (1956), *Les caractères originaux de l'histoire rurale française. 2, Supplément établi d'après les travaux de l'auteur (1931–1944)*, Paris, Librairie Armand Colin.

Bloch, Marc (1966), *French Rural History*, London, Routledge.

Bok, Derek (1982), *Beyond the Ivory Tower: Social Responsibilities of the Modern University*, Cambridge, Harvard University Press.

Bok, Derek (1986), *Higher Learning*, Cambridge, Harvard University Press.

Bok, Derek (2003). *Universities in the Marketplace: The Commercialization of Higher Education,* Princeton, N.J., Princeton University Press.

Bolin, Sture (1941), "Medieval Agrarian Society in its Prime 8. Scandinavia", in Clapham, John & Power, Eileen (eds.), *The Cambridge Economic History of Europe.*

Vol. 1, The Agrarian Life of the Middle Ages, Cambridge, Cambridge University Press.

Bolin, Sture (1952), *Sakkunnigutlåtande rörande de sökande till professuren i historia vid Uppsala universitet 1952,* Uppsala.

Bolin, Sture (1966), "Medieval Agrarian Society in its Prime 8. Scandinavia", in M.M. Postan (ed.), *The Cambridge Economic History of Europe. Vol. 1, The Agrarian Life of the Middle Ages*, Cambridge, Cambridge University Press.

Boserup, Ester (1999), *My Professional Life and Publications 1929–1998*, Copenhagen, Museum Tusculanum Press.

Bourdieu, Pierre (1977), *Outline of a Theory of Practice*, Cambridge, Cambridge University Press.

Bourdieu, Pierre (1988). *Homo Academicus,* Stanford, Stanford University Press.

Bramwell, Anna (1989), *Ecology in the 20th Century*, New Haven, Yale University Press.

Breisach, Ernst (1983), *Historiography: Ancient, Medieval, and Modern*, Chicago, University of Chicago Press.

Brichford, Maynard (1995), *A Brief History of the University of Illinois*, Urbana-Champaign.

Brooks, Paul (1989), *Rachel Carson. The Writer at Work*, San Francisco, Sierra Club Books.

Brown, Michael & May, John (1989), *The Greenpeace Story,* London, Dorling Kindersley Limited.

Burke, Peter (1990), *The French Historical Revolution: The Annales School, 1929–89*, Stanford, Stanford University Press.

Burke, Peter (1992), *History and Social Theory*, Cambridge, Polity.

Carlson, Benny (1995), *De institutionalistiska idéernas spridning*, Stockholm, SNS.

Carlsson, Sten (1956), *Bonden i svensk historia del 3*, Stockholm, Lantbruksförbundets tidskriftsaktiebolag.

Carson, Rachel (1970), *Silent Spring*, New York, Fawcett Press.

Chubin, Daryl E. & Hackett, Edward J. (1990), *Peerless Science: Peer Review and U.S. Science Policy*, New York, State University of New York press.

Clapham, John H. (1941), "Preface", in Clapham, John H. & Power, Eileen (eds.), *The Cambridge Economic History of Europe. Vol. 1, The Agrarian Life of the Middle Ages*, Cambridge, Cambridge University Press.

Clapham, John H. & Power, Eileen (eds.) (1941), *The Cambridge Economic History of Europe. Vol. 1, The Agrarian Life of the Middle Ages*, Cambridge, Cambridge University Press.

Clark, Burton R. (1987), *The Academic Life: Small Worlds, Different Worlds*, Princeton, Carnegie foundation for the advancement of teaching.

Chorafas, Dimitris N. (1968), *The Knowledge Revolution: an Analysis of the International Brain Market*, New York, McGraw-Hill.

Cochrane, Willard (1983), *Agricultural Economics at the University of Minnesota 1886–1979*, St. Paul, University of Minnesota, Agricultural Experiment Station.

Cochrane, Willard (1993), *The Development of American Agriculture. A Historical Analysis*, 2 ed., Minneapolis, University of Minnesota Press.

Conley, Kathleen (1998), "University of Illinois (Urbana-Champaign, Illinois, U.S.A.), in Summerfield, Carol & Summerfield, Mary (eds.), *International Dictionary of University Histories*, London, Fitzroy Dearborn, p. 507–510.

Deffeyes, Kenneth (2001), *Hubbert's Peak*, Princeton, N.J., Princeton University Press.

Donoghue, Frank (2008), *The Last Professors. The Corporate University and the Fate of the Humanities*, New York, Fordham University Press

Dorner, Peter (1972), *Land Reform and Economic Development*, Harmondsworth, Penguin Books.

Dorner, Peter (1992), *Latin American Land Reforms in Theory and Practice. A Retrospective Analysis*, Madison, The University of Wisconsin Press.

Dovring, Folke (1947), *Attungen och marklandet: studier över agrara förhållanden i medeltidens Sverige*, Lund, Gleerupska Universitetsbokhandeln.

Dovring, Folke (1947b), "'Gilzla oc grutha' i Smålandslagen", in *Arkiv för nordisk filologi 1947*, Lund, Gleerups.

Dovring, Folke (1951), *De stående skatterna på jord 1400–1600*, Lund, Gleerup.

Dovring, Folke (1952), *Oförrättat uppdrag?*, Lund, Gleerupska universitetsbokhandeln.

Dovring, Folke (1952b), "Le probléme du manse le Sundgau. Etude d´histoire agraire d´après les documents cadastraux de l´abbaye de Saint-Alban de Bâle", *L´Alsace et la Suisse à traverse les siècles* 1952.

Dovring, Folke (1953), "Agrarhistorisk forskning och svensk medeltidshistoria", *Historisk tidskrift* 1953.

Dovring, Folke (1953b), *Agrarhistorien*, Stockholm, Hugo Gebers.

Dovring, Folke (1959), "The share of agriculture in a growing population", *Monthly Bulletin of Agricultural Economics and Statistics* 1959.

Dovring, Folke (1960), *History as a Social Science: An Essay on the Nature and Purpose of Historical Studies*, The Hague, Nijhoff.

Dovring, Folke (1961), "The Opportunity to Multiply: Demographic Aspects of Modern Colonialism", *The Journal of Economic History* 1961.

Dovring, Folke (1962), "Peasantry, Land Use, and Change" Review of Walter Firery, *Comparative Studies in Society and History* 1962.

Dovring, Folke (1962b), "Flexibility and security in agrarian reform programs", in *Agrarian Reform & Economic Growth in Developing Countries. Papers from a Seminar on Research Perspectives and Problems*, Washington, United States Department of Agriculture.

Dovring, Folke & Larsen, Hans K. (1962c), *The Danish Farm system,* Illinois Agricultural Economics 1962.

Dovring, Folke & Scofield, H. (1963), *Farm Real Estate Sales in Illinois*, Department of Agricultural Economics, University of Illinois.

Dovring, Folke (1964), "The share of agriculture in a growing population", in Carl Eicher & Lawrence Witt (eds.), *Agriculture in Economic Development*, New York, McGraw-Hill.

Dovring, Folke (1964b), "Land ownership and tenure reforms", in *Illinois Agricultural Economics* Dec. 1964.

Dovring, Folke (1965), *Land and Labor in Europe in the Twentieth Century: A Comparative Survey of Recent Agrarian History: with a chapter on Land reform as a Propaganda Theme by Karin Dovring*, The Hague, Nijhoff.

Dovring, Folke (1965b), "The Transformation of European Agriculture", in Habakkuk, H. & Postan M. (eds.) *The Cambridge Economic History of Europe. Vol. 6, The Industrial Revolutions and After: Incomes, Population and Technological Change*, Cambridge, Cambridge University Press.

Dovring, Folke (1966), "The Conditions of Agricultural Growth: The Economics of Agrarian Change under Population Pressure", Review of Ester Boserup", *Journal of Economic History* 1966.

Dovring, Folke (1968), "Folke Dovring", in Debus Allen (ed.) *Who's who in Science 1968*, Chicago, Marquis-Who's who.

Dovring, Folke (1969), "The Principle of Acceleration: A Non-Dialectical Theory of Progress", *Comparative Studies in Society and History* 1969.

Dovring, Folke (1969b), "The Principle of Acceleration: A Non-Dialectical Theory of Progress: Counter-Comment", *Comparative Studies in Society and History* 1969.

Dovring, Folke (1969c), "Eighteenth-century changes in European agriculture: A comment", *Agricultural History* 1969.

Dovring, Folke & Dovring, Karin (1971), *The Optional Society: An Essay on Economic Choice and Bargains of Communication in an Affluent world*, The Hague, M. Nijhoff.

Dovring (1973), "Distribution of farm size and income. Analysis by exponential functions", in *Land Economics* 1973.

Dovring (1974), "Soybeans", in *Scientific American*, Feb. 1974.

Dovring, Folke (1975), *A National Economic Policy for the United States*, Champagne-Urbana, Folke Dovring.

Dovring, Folke (1976), "Party platforms blueprints for disaster", in *Illinois Business Review* Oct. 1976.

Dovring, Folke (1977), *Gross and Net Productivity. A Problem of Aggregation Level*, Department of Agricultural Economics, University of Illinois.

Dovring, Folke (1978), "The farm land boom in perspective. Recent history and economic relationships of farmland values", in *Effects of High Farm Land Prices,* Department of Agricultural Economics, University of Illinois.

Dovring, Folke (1978b), "Declare war on energy problem" in *The Morning Courier (Champagne-Urbana)* May 28 1978.

Dovring, Folke (1979), "Gross and net productivity. A problem of aggregation", in *American Journal of Agricultural Economics*, Vol. 61, No. 4.

Dovring, Folke (1981), Review of: Wilhelm Abel "Agricultural Fluctuations in Europe from the Thirteenth to the Twentieth Centuries", *Land Economics* 1981.

Dovring, Folke (1981b), "The oil crisis. A case of American underdevelopment", in *Illinois Business Review* April 1981.

Dovring, Folke, Braden, John & Chicoine, David, (1982), "Farmland conversion and preservation in Illinois", *Illinois CRD Report*, No. 2 March 1982.

Dovring, Folke (1982), "Oil Crisis and Vested Interests", in *Illinois Business Review*, April 1982.

Dovring, Folke (1984), *Riches to Rags*, Cambridge, Schenkman.

Dovring, Folke (1987), *Productivity and Value: The Political Economy of Measuring Progress*, New York, Praeger.

Dovring, Folke (1987b), *Land Economics*, Boston, Breton.

Dovring, Folke (1987c), *Productivity and Value: The Political Economy of Measuring Progress*, New York, Praeger.

Dovring, Folke (1988), *Farming for Fuel: The Political Economy of Energy Sources in the United States,* New York, Praeger.

Dovring, Folke (1988b), *Progress for Food or Food for Progress?: The Political Economy of Agricultural Growth and Development*, New York, Praeger.

Dovring, Folke (1991), *Inequality: The Political Economy of Income Distribution*, New York, Praeger.

Dovring, Folke (1996), *Leninism. Political Economy as Pseudoscience,* London, Praeger.

Dovring, Folke (1998), *Knowledge and Ignorance: Essays on Lights and Shadows*, London, Praeger.

Dovring, Karin (1995), "Maxim Fortuna", in *Faces in a Mirror*, Newport Beach, 1995.

Duby, Georges (1968), *Rural Economy and Country Life in the Medieval West*, London, Edward Arnold.

Dumont, René (1957), *Types of Rural Economy: Studies in World Agriculture*, London, Methuen.

Elgeskog, Valter (1945), *Svensk torpbebyggelse från 1500-talet till Laga skiftet. En agrarhistorisk studie*, Stockholm, LT.

Engberg, Magnus (2003). "En skånsk östgöte under svensk flagg", in *Från skilda utsikter. Skrivande om skrivande i Östergötland*, ed. Magnus Engberg, Stockholm, Arena.

Englund, Peter (2002), *Svenska Akademien inträdestal. Erik Lönnroth*, Stockholm, Svenska Akademien.

Erdmann, Karl Dietrich (1987), *Die Ökumene der Historiker: Geschichte der internationalen Historikerkongresse und des Comité international des sciences historiques*, Göttingen, Vandenhoeck & Ruprecht.

Erdmann, Karl Dietrich (2005), *Toward a Global Community of Historians. The International Historical Congresses and the International Committee of Historical Sciences 1898–2000*, New York, Berghahn Books.

Ericsson, Alf (2007), *Attungen – ett medeltida fastighetsmått*, Uppsala, Avdelningen för agrarhistoria, SLU.

Ericsson, Alf (ed.) (2008), *Jordvärderingssystem från medeltiden till 1600-talet*, Stockholm, Kungl. Vitterhetsakademien.

Forser, Tomas (1978), "En helig och allmännelig kyrka", in *Humaniora på undantag? Humanistiska forskningstraditioner i Sverige*, Stockholm, Norstedts.

Furuland, Lars (1989), "Folkrörelser och arbetardikt ca 1880–1920", in Lönnroth, Lars & Delblanc, Sven (eds.), *Den svenska litteraturen 4. Den storsvenska generationen 1890–1920*, Stockholm, Bonnier.

Gardner, Bruce L. (2002), *American Agriculture in the Twentieth Century*, Cambridge, Mass., Harvard University Press.

Gasson, Ruth & Errington, Andrew (1993), *The Farm Family Business*, Wallingford, CAB International.

Geiger, Roger, (1991), "The American university and research: A historical perspective", in Grigg, David B (ed.), *University and Society 1991, The Agricultural Systems of the World: An Evolutionary Approach*, London, Cambridge.

Geiger, Roger (1993), *American Research Universities since World War II*, Oxford, Oxford University Press.

Glassick, Charles E., Huber, Mary Taylor & Maeroff, Gene I. (1997), *Scholarship Assessed: Evaluation of the Professoriate*, San Francisco, Jossey-Bass.

Graham, Hugh Davis & Diamond, Nancy (1997), *The Rise of American Research Universities. Elites and Challengers in the Postwar Era*, Baltimore, John Hopkins University press.

Grigg, David B. (1974), *The Agricultural Systems of the World: An Evolutionary Approach*, London, Cambridge.

Gunneriusson, Håkan (2002), *Det historiska fältet: svensk historievetenskap från 1920-tal till 1957*, Uppsala, Acta Universitatis Upsaliensis.

Halcrow, Harold (1998), *The Laws of My Life*, H.G. Halcrow.

Hankins, Thomas (1979), "In defence of biography: The use of biography in the history of science", in *History of Science* 1979.

Hayami, Yujiro & Ruttan, Vernon W. (1985), *Agricultural Development: An International Perspective*, Baltimore, Johns Hopkins University Press.

Hasselberg, Ylva (2007), *Industrisamhällets förkunnare. Eli Heckscher, Arthur Montgomery, Bertil Boëthius och svensk ekonomisk historia 1920–1950*, Hedemora, Gidlunds förlag.

Hedman, Dag (1994), "Ossiannilsson, Karl Gustav", in *Svenskt biografiskt lexikon* Vol. 28 (Odeberg–Pederby), Stockholm, SBL.

Hughes, Donald (2001), *Världens miljöhistoria*, Kristianstad, SNS.

Husberg, Erik (1994), *Honung, vax och mjöd: biodlingen i Sverige under medeltid och 1500-tal*, Göteborg, Historiska institutionen i Göteborg.

Iggers, Georg (2005), *Historiography in the Twentieth Century*, Middletown, Wesleyan University Press.

Ingers, Enoch (1943–1948), *Bonden i svensk historia 1–2*, Stockholm, LT.

Jacoby, Eric (1971), *Man and Land. The Fundamental Issue in Development*, London, André Deutsch.

Johnston, Bruce F. & Kilby, Peter (1975), *Agriculture and Structural Transformation: Economic Strategies in Late-developing Countries*, New York, Oxford University Press.

Jansén, Jan (2005), "Hur hållbara var några olika historiska odlingssystem?", in *Bruka, odla, hävda. Odlingssystem och uthålligt jordbruk under 400 år*, (Skogs- och Lantbrukshistoriska meddelanden nr 33) Stockholm, KSLA.

Jencks, Christopher & Riesman, David (1969), *The Academic Revolution*, Garden City NY, Doubleday.

Jones, Eric Lionel (1988), *Growth Recurring: Economic Change in World History*, Oxford, Clarendon.

Kerr, Clark [et al.] (1978), "Comparative Effectiveness of Systems", in *12 Systems of Higher Education: 6 decisive issues*, New York, International Council for Educational Development.

Kerr, Clark (1982), *The Uses of the University*, Cambridge, Harvard University Press.

Kirp, David (2003), *Shakespeare, Einstein and the Bottom Line. The Marketing of Higher Education*, Cambridge, Harvard University Press.

Kimbrell, Andrew (2002) (ed.), *The Fatal Harvest Reader. The Tradegy of Industrial Agriculture,* Washington, D.C., Island Press.

King, Russell (1977), *Land Reform – A World Survey*, Boulder, Westview Press.

Kuhn, Thomas (1969), "The Principle of Acceleration: A Non-Dialectical Theory of Progress: Comment", *Comparative Studies in Society and History* 1969.

Kurian, George Thomas (1997), *The Illustrated Book of World Rankings*, Armonk N.Y., Sharpe Reference.

Landgren, Bengt (2005), "Universitetsämne i brytningstid. Svensk akademisk litteraturundervisning 1947–1956", in *Universitetsämne i brytningstider. Studier i svensk akademisk litteraturundervisning 1947–1995*, Landgren, B. (ed.), Uppsala, Uppsala universitet.

Lindstedt, Gunnar (2006), *Olja*, Stockholm, Månpocket.

Linnér, Björn-Ola (1998), *The World Household: Georg Borgström and the Postwar Population-resource Crisis*, Linköping, Tema Univ.

Linnér, Björn-Ola (2003), *The Return of Malthus. Environmentalism and Post-war Population-Resource Crises*, Isle of Harris, The White Horse Press.

Lönnroth, Erik (1963), "The Baltic Countries", in Postan, M.M., Rich E.E., & Miller Edward (eds.) *The Cambridge Economic History of Europe, Volume 3, Economic Organization and Policies in the Middle Ages*, Cambridge, Cambridge University Press.

Lönnroth, Erik (1998), *Tidens flykt. Stora historiska förändringar och människor som har levat i dem,* Stockholm, Atlantis.

Markland, *Nationalencyklopedin*, Vol. 13.

Martling, Göran (2002), "Om trummande skalbaggar. Internet och omslagsvinjetter", *Biblis* 19/20 2002.

McConnell, Grant (1954), "The Conservation Movement, Past and Present", in *The Western Political Quarterly* 1954.

Mellor, John W. (1966), *The Economics of Agricultural Development*, Ithaca, N.Y., Cornell University Press.

Moores, Richard Gordon (1970), *Fields of Rich Toil. The Development of The University of Illinois College of Agriculture*, Urbana, University of Illinois Press.

Muir, John (1997), *Nature Writings*, New York, Library of America.

Myrdal, Gunnar (1968), *Asian Drama: An Inquiry into the Poverty of Nations*, New York, Pantheon.

Myrdal, Janken (1982), "August Strindberg och den stora striden om kulturhistoria 1881–82", in *En annan historia: sju folkliga historiker från Nils Månsson Mandelgren till Vilhelm Moberg*, Stockholm, Ordfront.

Myrdal, Janken (2001), "Peasant and rural societies in history, agricultural history", in *International Encyclopedia of the Social & Behavioral Sciences*, Vol. 16, Amsterdam, Elsevier.

Myrdal, Janken (2003), *Digerdöden, pestvågor och ödeläggelse*, Stockholm, Sällskapet Runica et Mediævalia.

Myrdal, Janken & Söderlind, Ingrid (2003), "De disputerade är en resurs: Sverige går om USA i avhandlingsproduktion", *Dagens forskning* May 12–13 2003.

Myrdal, Janken (2008), "Förorden i forskningsfältet – historiker i mitten av 1900-talet", in *Kritikens dimensioner. Festskrift till Tomas Forser*, Stockholm, Brutus Östlings Bokförlag.

Myrdal, Janken (2008b), "De medeltida jordvärderingarnas historiografi", in *Jordvärderingssystem från medeltiden till 1600-talet*, Stockholm, Kungl. Vitterhetsakademien.

Nash, Roderick (1967), "Gifford Pinchot", in Pinchot, Gifford *The Fight for Conservation*, Seattle, University of Washington Press.

Niemi, Albert (1975), "Journal Publication Performance of Economic History Programs: 1960–1969 and 1970–1974", *Journal of Economic History* 1975.

Nilsson, Bo G. (2004), "Framtidens salt. Om museernas och folklivsforskarnas bidrag till folkhemsbygget", in *Samhällsideal och framtidsbilder*. Stockholm, Carlssons.

Nilsson, Sven (1952), *Självtagen sakkunskap: Erinringar i anledning av docenterna Sten Carlssons och Folke Dovrings inlagor rörande den ledigförklarande professuren i historia vid Uppsala universitet*, Lund, Gleerupska universitetsbokhandeln.

Norborg, Lars-Arne (1958), *Storföretaget Vadstena kloster: studier i senmedeltida godspolitik och ekonomiförvaltning*, Lund, Gleerup.

Odén, Birgitta (1975), *Lauritz Weibull och forskarsamhället*, (Bibliotheca historica Lundensis), Lund, Liber Läromedel/Gleerup.

Odén, Birgitta (1978), "Annales-skolan och det svenska forskarsamhället", in *Att skriva historia. Nya infallssvinklar och objekt*, Stockholm, Norstedts.

Odén, Birgitta (1991), *Forskarutbildningens förändringar 1890–1975*, Lund, Lund University Press.

Pachter, Marc (1979), "The Biographer Himself: An Introduction", in Pachter, Mark (ed.) *Telling Lives – The Biographer's Art*, Washington D.C., New Republic Books.

Palme, Sven Ulric (1953), "Agrarhistorien av Folke Dovring", *Stockholmstidningen* August 24 1953.

Perlinge, Anders (manuscript), *Börje Hanssen 1917–1979.*

Pinchot, Gifford (1967), *The Fight for Conservation*, Seattle, University of Washington Press.

Poly, Jean-Pierre & Bournazel, Eric (1991), *The feudal transformation 900–1200*, New York, Holmes & Meier.

Postan, Michael Moïssey (1972), *The Medieval Economy and Society: An Economic History of Britain in the Middle Ages*, London, *Weidenfeld & Nicolson*.

Putnam, Robert (2000), *Bowling Alone: The Collapse and Revival of American Community*, New York, Simon & Schuster.

Raphael, Lutz (2003), *Geschichtswissenschaft im Zeitalter der Extreme. Theorien, Methoden, Tendenzen von 1900 bis zur Gegenwart*, München, Beck.

Reiss, Franklin J. (1982), *Milestones and memories from the firsts fifty years of the Department of Agricultural Economics*, Urbana-Champaign, University of Illinois.

Rosovsky, Henry (1990), *The University: An Owner's Manual*, New York, W.W. Norton.

Rösener, Werner (1992), *Peasants in the Middle Ages,* Cambridge, Polity Press.

Sandelin, Bo, Trautwein, Hans-Michael & Wundrak, Richard (2001), *Det ekonomiska tänkandets historia*, Kristianstad, SNS.

Sandemose, Jørgen (2005), *Biografi og metode: en pamflett til filosofisk og litteraer oppbyggelse*, Oslo.

Shatz, David (2004), *Peer Review: a Critical Inquiry*, Lanham, Maryland, Rowman & Littlefield Publ.

Schück, Herman (1980), "Centralorgan för den svenska historiska forskningen. Historisk tidskrift från sekelskiftet till 1960-talets början", *Historisk tidskrift* 1980.

Schück, Herman (1994), *Engelbrektskrönikan; tillkomsten och författaren*, Stockholm, Kungliga Vitterhets-, historie- och antikvitetsakademiens handlingar, Historiska serien.

Servan-Schreiber, Jean Jacques (1968), *The American challenge*, Harmondsworth, Penguin.

Schuster, J.H. (2002), "Academic labor market", in Clark, B.R. & Neave, G. (eds.), *Encyclopedia of Higher Education*, Vol. 3, Oxford, Pergamon Press.

Simerl, L.H. (1961), "Has Denmark saves or destroyed the family farm?", *Illinois Farmer's Outlook Letter*, Urbana, 20/12 1961, (mimeographed, 2p).

Stadtman, Verne (1991), "United States", in Clark, B.R.& Neave, G. (eds.), *The Encyclopedia of Higher Education*, Vol. 1, Oxford, Pergamon Press.

Steen, Sverre (1952), *Sakkunnigutlåtande rörande de sökande till professuren i historia vid Uppsala universitet 1952,* Uppsala.

Strange, Marty (1988), *Family Farming: A New Economic Vision*, The Institute for Food and Development, Oakland, CA.

Sundquist, Burt W. (2001), *A History of Agricultural and Applied Economica at the University of Minnesota 1979–2000*, Minnesota: Department of Applied Economica, University of Minnesota.

Svensson, Sven G. (1994), *Erland Hjärne. En historikers livsöde*, Stockholm, Natur och Kultur.

Therborn, Göran (1995), *European Modernity and Beyond. The Trajectory of European Societies 1945–2000*, London, Sage Publications.

Thoen, Erik & Van Molle, Leen (2006) (eds.) *Rural History in the North Sea Area. An Overview of Recent Research (Middle Ages–Twentieth Century)*, Brepols, Turnhout.

Thrupp, Sylvia (1964), "Some historians on generalization", *Comparative Studies in Society and History* 1964.

Torstendahl, Rolf (1964), *Källkritik och vetenskapssyn i svensk historisk forskning 1820–1920*, Stockholm, Svenska bokförlaget Norstedt.

Torstendahl, Rolf (1971), *Introduktion till historieforskningen: historia som vetenskap*, Stockholm, Natur och Kultur.

Torstendahl, Rolf & Nybom, Thorsten (1988), *Historievetenskap som teori, praktik, ideologi*, Stockholm, Författarförlaget.

Trow, Martin (1989), "American Higher Education: Past, Present, and Future", in Goodchild, Lester F. & Wechsler, Harold S. (eds.) *The History of Higher Education, ASHE Reader Series*, Needham Heights, Ginn Press.

Trow, Martin (1991), "The Exceptionalism of American Higher Education", in Trow, Martin & Nybom, Thorsten (eds.) *University and Society: Essays on the Social Role of Research and Higher Education*, London, Jessica Kingsley.

Tuchman, Barbara (1979), "Biography as a Prism of History", in Pachter, Marc (ed.) *Telling Lives – The Biographer's Art*, Washington D.C., New Republic Books.

Utterström, Gustaf (1955), "Book review: Agrarhistorien. En översikt av dess uppgifter, forskningsmetoder och resultat", *Historisk tidskrift* 1955.

Utterström, Gustaf (1957), *Jordbrukets arbetare: levnadsvillkor och arbetsliv på landsbygden från frihetstiden till mitten av 1800-talet*, Stockholm, Tiden.

Vasey, David (1992), *An Ecological History of Agriculture: 10,000 B.C.–A.D. 10,000*, Ames, Iowa State University Press.

Verhulst, Adriaan (2002), *The Carolingian Economy*, Cambridge University Press, Cambridge.

Wallerstein, Immanuel (1989), The *Modern World-system. 3, The Second Era of Great Expansion of the Capitalist World-economy, 1730– 1840s*, New York, Academic Press.

Weibull, Curt (1951–52), "En ny generation svenska historiker", *Scandia: Tidskrift för historisk forskning.*

Whürer, Karl (1957), "Die agrargeschichte Forschung in Skandinavien, seit 1945", *Zeitschrift für Agrargeschichte und Agrarsoziologie* 1957.

Wittfogel, Karl A., (1980), *Oriental Despotism: A Comparative Study of Total Power*, New York, Vintage Books.

Österberg, Eva (1970), "Skatter", in *Kulturhistoriskt lexikon*, Vol. 15, Malmö, Allhems förlag.

Private Archive of Maj Sundblad (PA-MS)

Collection of family letters.

Private Archive of Philip Raup (PA-PR)

Letters to and from Folke Dovring.

The Folke Dovring Archive, University of Illinois (DA)

Letters to and from Folke Dovring.

Course and Instructor Evaluation, Fall Semester 1971.

MS Examination of V.V. Sharma, 27.8.1965.

Working List for Agricultural Economics Staff and Graduate Students, 15.6.1960.

Report to The Rockefeller Foundation on work under Social Sciences fellowship, November 1952, to October 31 1954.

Utdrag av Humanistiska sektionens i Uppsala protokoll den 1 oktober 1952. (Copy from Archive of the University of Uppsala.)

Sakkunnigutlåtanden rörande tillsättandet av professuren i historia vid Göteborgs högskola 1953. (Copy from Archive of the University of Göteborg.)

Utdrag ur protokoll, hållet hos Humanistiska Sektionen i Lund den 28 februari 1947. (Copy from Archive of the University of Lund.)

Utdrag ur protokoll, hållet vid sammanträde med juridiska fakulteten vid Lunds universitet den 21 mars 1947. (Copy from Archive of the University of Lund.)

Archive and Library of the Labor Movement (Arbetarrörelsens arkiv och bibliotek), Stockholm

Alva och Gunnar Myrdals arkiv, Vol. 3.2.2: Korrespondens; Gunnar Myrdal: UN Economic Comission for europé (MA 3.2.2); Vol. 3.2.3: Korrespondens Gunnar Myrdal (MA 3.2.3)

National Heritage Board, Department for Archaeological Excavations: East (Linköping, Östergötland), Office library

Gunnar Westin's copy of Folke Dovring: *Attungen och marklandet*

Copies of most of the letters and other documents, also in KSLAB – Kungl Skogs och Lantbruksakademiens bibliotek, Stockholm.

Interviews

Bloomkvist, Tim: June 17, 2002.

Bunker, Stephen: May 21, 2002.

Chicoine, David: June 18, 2002.

Dovring, Karin; several times during 2002.

Fischer, Ralph: May 23, 2002.

Harcrowe, Harold: May 24, 2002.

Husén, Torsten: March 25, 2002.

Kellog, Earl: July 11, 2002.

Odén, Birgitta, May 9, 2003.

Petersson, Jan: March 23, 2002.

Raup, Philip: June 27, 2002.

Spitze, Robert: Several times during 2002

Sundblad, Maj: March 30, 2002.

Swanson, Earl: Several times during 2002.

Westergren, Randall: June 20, 2002.

Westin, Gunnar: February 3, 2002 and June 6, 2002.

KVHAA HANDLINGAR *Historiska serien*

1 Boëthius, B., *Dalarnas bränsleskatter och Stora Kopparbergs bergsfrälse under äldre Vasatid.* 1957

2 Viljanti, A., *Gustav Vasas ryska krig 1554–1557.* 1–2 (The Russian War of Gustav Vasa 1554–1557). 1957

3 Axelson, S., *Sverige i dansk annalistik 900–1400.* 1956

4 Olofsson, S.I., *Efter Westfaliska freden. Sveriges yttre politik 1650–1654* (Nach dem Wästfälischen Frieden. Die Aussenpolitik Schwedens 1650–1654). 1957

5 Odén, B., *Kopparhandel och statsmonopol. Studier i svensk handelshistoria under senare 1500-talet* (Kupferhandel und Staatsmonopol. Studien in schwedischer Handelsgeschichte im späteren 16. Jahrhundert). 1960

6 Carlsson, G., *Från Erik Segersäll till Gustav Vasa. Undersökningar och rön* (Von Erik Segersäll bis Gustav Vasa. Untersuchungen und Ergebnisse). 1961

7 Fahlborg, B., *Sveriges yttre politik 1668–1672.* 1–2. 1961

8 Soom, A., *Der Baltische Getreidehandel im 17. Jahrhundert.* 1961

9 Steckzén, B., *Birkarlar och lappar. En studie i birkarleväsendets, lappbefolkningens och skinnhandelns historia.* 1964

10 *Gustavianska opinioner. Michael Anckarsvärds brev till Nathanael Gerhard Schulthén 1790–1808.* Med inledning och kommentarer utgivna av Mustelin, O. 1965

11 Klockars, B., *Birgitta och böckerna. En undersökning av den heliga Birgittas källor* (Sources of the writings of St. Bridget of Sweden). 1966

12 Ekwall, S., *Vår äldsta Birgittavita och dennas viktigaste varianter* (La plus ancienne vie de S:te Brigitte et ses deux variations les plus importantes). 1965

13 Ström, A., *Huvuddragen av Olof Håkanssons politiska verksamhet åren 1726–1743* (The political activities of Olof Håkansson, 1726–1743). 1967

14 Bachman, M.-L., *Studier i Kungl. Vitterhets Historie och Antikvitets Akademiens historia* (History of the Royal Swedish Academy of Letters, History and Antiquities, 1837–1879). 1969

15 Soom, A., *Die Zunfthandwerker in Reval im siebzehnten Jahrhundert.* 1971

16 Klockars, B., *Birgitta och hennes värld.* 1971

17 Friedlaender, I., *Peringskölds Diplomatarium.* (Zusammenfassung) 1971. ISBN 91-7192-004-8

18 Lundberg, B., *Territoriell indelning och skatt i Uppland under medeltiden* (Territoriale Einteilung und Besteuerung in Uppland im Mittelalter). 1972. ISBN 91-7192-045-5

19 Arnell, S., *Bidrag till belysning av den baltiska fronten under det nordiska sju-årskriget 1563–1570*. 1977. ISBN 91-7402-033-1

20 Kumlien, K., *Historieskrivning och kungadöme i svensk medeltid*. 1979. ISBN 91-7402-094-3

21 Klockars, B., *I Nådens dal. Klosterfolk och andra c. 1440–1590*. 1979. ISBN 91-7402-079-X

22 Schück, H., *Engelbrektskrönikan. Tillkomsten och författaren*. 1994. ISBN 91-7402-245-8

23 Schück, H., *Rikets råd och män. Herredag och råd i Sverige 1280–1480*. 2005. ISBN 91-7402-352-7